William Stukeley Gresley

**A Glossary of Terms Used in Coal Mining**

William Stukeley Gresley

**A Glossary of Terms Used in Coal Mining**

ISBN/EAN: 9783744790888

Printed in Europe, USA, Canada, Australia, Japan

Cover: Foto ©Andreas Hilbeck / pixelio.de

More available books at **www.hansebooks.com**

A

# GLOSSARY

OF

## TERMS USED IN COAL MINING.

BY

WILLIAM STUKELEY GRESLEY,

ASSOC. MEM. INST. CIVIL ENGINEERS, F.G.S., MEM. NORTH OF ENGLAND INSTITUTE OF
MINING AND MECHANICAL ENGINEERS, ETC. ETC. ETC.

ILLUSTRATED WITH NUMEROUS WOODCUTS AND DIAGRAMS.

LONDON:
E. & F. N. SPON, 16, CHARING CROSS.
NEW YORK: 35, MURRAY STREET.

1883.

## Dedicated

WITH SINCERE AFFECTION AND RESPECT TO THE

MEMORY OF THE LATE

## JOHN THOMAS WOODHOUSE, Esq.,

M.INST.C.E., F.G.S., M.N.E.I.M.E., M.I. AND S.I., M.S.E., A.I.N.A., ETC.

# PREFACE.

THE Compiler of the following Glossary of Terms used in connection with the mining of coal and other minerals had at first no intention of publishing his work. He merely collected, arranged, and classified the various local and provincial mining terms and phrases as they came under his notice, for his own personal curiosity and use. At the request of several friends, however, he has decided to go more minutely and carefully into the subject, and has made an attempt to give to the mining community, and others interested in the science of coal mining, the result of a much closer investigation into the study of the provincialisms and technicalities of the mining districts of this country; and although conscious of its many defects, he now ventures to offer to the scientific public the accompanying compendium of the terms employed in the mining of coal and other stratified minerals.

It is also hoped that many of the terms have been explained in such a manner as not only to give a meaning, as clear and concise as is possible under the circumstances, but also to convey to students and others engaged in the mining profession some information, in detail, as to the several methods, operations, systems, appliances, statistics, &c., used in connection with the winning, working, and disposal of Coal, which has so often been described as the " Mainspring of Civilisation," and which, owing to the bountiful munificence of the Creator and Giver of all good things, has made Great Britain what she is, viz. by far the largest producer, hitherto, of that mineral in the world.

OVERSEAL, ASHBY DE LA ZOUCH,
*December*, 1882.

# INTRODUCTION.

In introducing the reader to the contents of this little work, it may be well in the first place to give some explanation of the method adopted in compiling it, and to mention some of the sources from whence many of the words and phrases have been obtained.

As many of the terms treated of have been gathered from journals, reports, and transactions of mining institutes, &c., it is not improbable that several inaccuracies may be met with, the meaning given not being in all cases so explicit as the Compiler could have wished; but by the exercise of much care and considerable labour, he believes that they have been reduced to a very few.

Any one who will be good enough to favour the Compiler with terms, &c., omitted or hitherto unknown to him, or with corrected and more accurate information, will be greatly assisting to improve, complete, and enhance the value of a subsequent edition, should it be called for.

It has been thought well to insert many terms which now are or are rapidly becoming obsolete, because it seemed unnecessary and impossible to draw a hard and fast line between them—obsolete words being interesting if not instructive to many.

To some terms a historical fact or addition has been included, by way of imparting information to the uninitiated.

As to words made use of in great number and variety in reference to Strata, or the names given to various beds of rock met with in the course of mining, these are so intimately mixed up with many of the terms used underground, that to exclude them would have been unfair.

With reference to the fact that very many terms have more than one, in some instances eight or ten, separate meanings, and that a single article, &c., may have as many as twelve or fourteen different names by which it is called, it must be understood that the numbers (1, 2, 3, &c.) placed immediately after a word refer to corresponding numbers under the head of which the explanation of the particular term will be found, e. g. "The *box* at the *head* (1) *end* has only one *garland* (2) upon it." By looking out the word *head* under No. 1 explanation, and *garland* under No. 2 meaning, will at once give the reader an idea of the system upon which the whole book is drawn up.

Again, with regard to machinery and mechanical appliances generally, it has been thought proper to exclude all technical terms applied to the various parts of such things as do not refer especially to mining, for instance:— the words *pump, boiler, donkey, fly-*

*wheel, points, spanner, cotter,* &c., are none of them included.

A number of terms have been obtained from the coal districts of Pennsylvania and elsewhere in America, but some of them are clearly traceable to the north of England, whence doubtless they originally came. Many Belgian, French, Prussian, German, Italian, &c., terms have been inserted, it being thought the better plan to leave out nothing that might in any way contribute to the usefulness of the work.

Turning to the sources of information of which the compiler has been so far able to avail himself, he hereby desires to acknowledge his thanks to various authors for giving many technical and local terms, in their various papers, addresses, books, and so forth, which he has ventured to make use of. The figures accompanying the text have, with only one or two exceptions, been drawn up by the writer expressly for the work, and he only regrets that this portion of his labours has been so imperfectly performed. The following are the principal works and authors consulted: The Transactions of the North of England Institute of Mining and Mechanical Engineers; the Proceedings of the South Wales Institute of Engineers; the Transactions of the Chesterfield and Derbyshire Institute of Engineers; the Transactions of the Mining Institute of Scotland; the Transactions of the Manchester Geological Society; the Transactions of the Midland Institute of Mining, Civil, and Mechanical

*b*

Engineers; the Annual Reports of H. M. Inspectors of Mines; the *Colliery Guardian* newspaper; Mine Engineering, by G. C. Greenwell; Mine Engineering, by G. E. André; the Journal of the British Society of Mining Students; as well as numerous smaller works chiefly relating to coal mining. It should, however, be remarked that the compiler has himself, in the course of his professional duties, visited nearly all the coal-fields of Great Britain, thus enabling him to acquaint himself pretty well with many of the terms commonly made use of. To compile a *complete* glossary of such terms would, it is believed, occupy many years, even if it were possible to do it at all.

In conclusion it should be said, that besides the terms and phrases used in coal mining, those used in connection with the working of ironstone, shale, fireclay, rock-salt, stone, &c.—in short, *stratified mines*, have been freely dealt with.

# ABBREVIATIONS.

B.     Bristol Coal-field.
Belg.    Belgium.
C.     Cumberland Coal-field.
Ch.     Cheshire Salt Districts.
Cl.     Cleveland Iron Districts.
D.     Derbyshire Coal-field.
F.     France.
F. D.   Forest of Dean Coal-field.
G.     Gloucestershire Coal-field.
I.     Ireland.
In.     India.
It.     Italy.
L.     Lancashire Coal-field.
Lei.    Leicestershire Coal-field.
M.     Midland Coal-field.
N.     North of England (Northumberland and Durham).
N. S.   North Staffordshire.
N. S. W. New South Wales.
N. W.   North Wales.
Pa.     Pennsylvania, U. S. A.
Pr.     Prussia.
S.     Scotland.
Sh.     Shropshire.
Som.   Somersetshire.
S. S.   South Staffordshire.
S. W.   South Wales.
Sw.    Sweden.
U. S. A. United States of North America.
W.     Warwickshire.
Y.     Yorkshire.

A

# GLOSSARY

OF

## TERMS USED IN COAL MINING, &c.

### A.

ABATTIS (Lei.). Walls or ranges of branch or rough wood (cord-wood) placed crossways to keep the underground roads open for ventilation, &c.

ABTHEILUNG (Pr.). A fixed part or *district* of a mine assigned to the care of a *fire-man* or *deputy*.

ACREAGE RENT. Royalty or rent paid by the lessee for working and disposing of minerals at the rate of so much per acre. Very frequently this rent is calculated at so much per foot thick of the seam or mine per acre, the measurements being taken on the slope or plane of the coal, &c., and at right angles to the *dip*.

ADAMANT (N. W.).

ADDLE (N.). To earn.

ADDLINGS (N.). Earnings or wages.

ADIT. An underground level to the surface from the level of the mine workings, or from part of the way down the shaft (Fig. 1), generally used for drainage purposes.

Fig. 1.

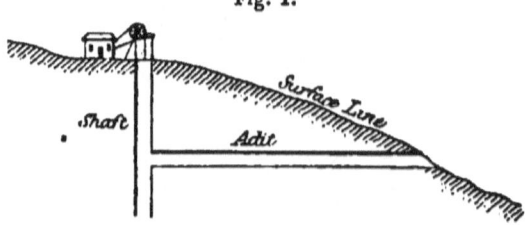

ADVENTURERS. The original promoters or speculators in a search for coal, &c.

AEROMETERS. The air pistons of a Struvé *ventilator*.

AEROPHORE. The name given to an apparatus which will enable a man to enter places in mines filled with explosive or other deadly gases, work there with freedom, take with him a light, and remain for an indefinite time.

AFTER-DAMP. The deadly gases resulting from an explosion of *fire-damp*. Chiefly composed of carbonic acid gas. $CO_2$ or carbon 27 per cent. + oxygen 73 per cent.

AGENT. One to whom the general laying out and supervision of the *workings* is entrusted by the owner or lessee. He may have a number of separate collieries under his care. The wages and contractor's prices are regulated by him. Any addition or alteration in the various departments connected both with the underground and surface works, machinery, &c., must generally be sanctioned by him. He is responsible to the owner as well as under the Coal Mines Regulation Act for the appointment of competent *managers, enginewrights, deputies*, surveyors, &c. See *Viewer*.

AIR. 1. The current of atmospheric air circulating through and ventilating the *workings* of a mine.
2. To ventilate any portion of the *workings*.

AIR-BOX. A rectangular wooden pipe or tube made in lengths of say 9 to 15 feet for ventilating a *heading* or a *sinking pit*.

AIR-COURSE. Any underground roadway used for the special purpose of ventilation.

AIR-CROSSING. A bridge which carries one *air-course* over another. In collieries liable to heavy explosions, in order to prevent as far as possible the *blast* from destroying these air-crossings and deranging the ventilation, it is better to avoid the use of the ordinary

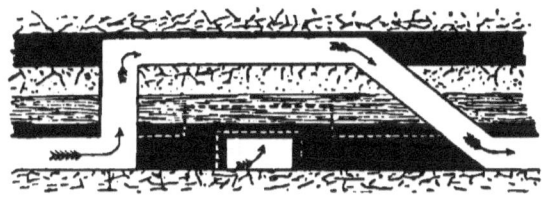

Fig. 2.

timber or even masonry bridge, and to make an entirely isolated air-course several yards above the underneath road, and if a seam of coal be conveniently situated in which to construct it, it will not be an expensive plan. See Fig. 2. (The dotted lines show the position of an ordinary crossing.)

AIR-END WAY. Roadways or levels driven in the coal seam parallel with a main level, chiefly for the purpose of ventilation or for the *return air*. They are connected with the main level by *openings* or *thirls*.

AIR-GATES (M.). Underground roadways used principally for ventilative purposes.

AIR-HEAD. See *Air-way*.

AIRLESS END. The extremity of a *stall* in *long-wall* workings in which there is no current of air, or circulation of ventilation, but which is kept *sweet* by diffusion, and by the ingress and egress of *tubs*, men, &c.

AIR-LEVEL. A level or *air-way* (return air-way) of former workings, made use of in subsequent deeper mining operations for ventilating purposes.

AIR-PIT. A *pit-shaft* used expressly for *ventilation*.

AIR-SLIT (Y.). A short *head* (1) driven more or less at right angles to, and between other two *heads* or *levels* for ventilation purposes.

Fig. 3.

AIR-SOLLAR. A *brattice* carried beneath the tram-rails in a *heading*, a, Fig. 3.

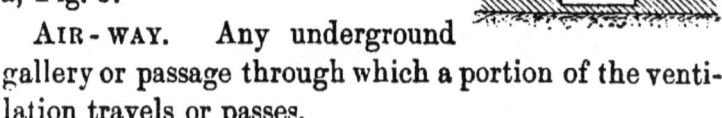

AIR-WAY. Any underground gallery or passage through which a portion of the ventilation travels or passes.

ALLOWANCE. 1. Refreshment in the shape of bread and cheese and beer supplied by the lessees or owners of a mine to surveyors who *dial* the *workings* periodically.

2. Ale sometimes given to workmen on having to perform work under unusual conditions, e. g. when they are wet through.

ALLOWANCE COAL. See *Colliers' coals*.

ALL-UPS (Lei.). A mixture of every quality of coal, excepting fine slack, raised from one *seam*, and sold as such.

ALL WORK (D.). Term formerly used for *Longwall*, which see.

ALTOGETHER-COAL. Large and small mixed.

ALUM SHALE. Earth containing the mineral alum, beds of which occasionally occur in the coal measures, sometimes as an *underclay*.

ANEMOMETER. An instrument for measuring the velocity of the ventilating current in mines.

ANTHRACITE. A hard, clean, bright, smokeless, and very pure variety of coal, having a conchoidal fracture, and burning with little or no flame, but containing very great local heating properties. It is much esteemed for malting and steam raising. It frequently contains over 90 per cent. of carbon; some of the anthracites of Pembrokeshire contain as much as 94 per cent. This coal weighs from 85 to 99·5 lbs. per cubic foot.

APPARATUS (N.). The screening appliances upon the pit bank.

ARCHING. Brickwork or stonework forming the roof of any underground roadway.

ARLES OR EARLES (N.). Earnest money formerly allowed to colliers at the time of hiring them.

ASCENSIONAL VENTILATION. The arrangement of the ventilating currents in such-wise that the heated air shall continuously rise until reaching the bottom of the upcast shaft. Particularly applicable to steep seams or *rearers*.

ASH-BALL (Sh.). Mixed small fragments of greenish clay, quartz, &c.

ATTLE (N.). To arrange or settle.

AUGER-NOSE SHELL. A clearing tool used in boring for coal, &c., having an auger-shaped end.

AVERAGE CLAUSE. One which, in granting leases of minerals (coal, ironstone, and clay in particular), provides that lessees may, during (say) every year of the term, make up any deficiency in the quantity of coal, &c., stipulated to be worked, so as to balance the dead or *minimum rent*.

AWARD (F. D.). A grant or lease of certain minerals. See *Gale*.

## B.

BACK. 1. A plane of cleavage in coal, &c., having frequently a smooth parting and some sooty coal included in it.

2. The inner end of a *heading* where work is going forward or is stopped.

3. (Lei.) To throw back into the *gob* or *waste*, the small slack, dirt, &c., made in *holing*.

4. (Lei.) To roll large coals out of a *waste* for loading into trams.

BACK-BOARD (Y.). A *thirl* communicating with the *return* air-course often fitted with a *regulator*.

BACK-BYE (N.). Work performed underground by the *deputies* after examining their districts in the pit, in *drawing* timbers in abandoned or worked-out places, repairing *brattices*, *doors*, &c., and attending and keeping in order the roadways, &c.

BACK-CASING. A wall or lining of dry bricks used in sinking through drift deposits, the permanent walling being built up within it. In the north of England the use of timber *cribs* and planking serves the same purpose.

BACK-COMING (S.). Working away the *pillars* left in, when getting coal *inbye*.

BACKEN (S. S.). See *Back* (4).

BACK-END (N.). A portion of a *jud*.

BACKING-DEALS. Deal boards or planking placed at the back of *curbs* for supporting the sides of a shaft liable to *run* (7).

BACK-LASH. The return or counter *blast* (1); recoil or backward suction of the air-current produced after an *explosion* of *fire-damp*.

BACK-LYE (S.). A siding or shunt on an underground tramway.

BACK-OVERMAN (N.). A man whose duty it is to see to the safety of a *district* of underground workings, and of the men working in it during the *back-shift*.

BACK-SHIFT (N.). A second shift or relay of *hewers* in each day, usually commencing work a few hours after the *drawing* (3) of coals begins.

BACK-SPLINTING (S.). A system of working a seam of coal over the *goaf* and across the *packs* of a lower one got in advance upon the *long-wall* method. *Back-splinting* consists in taking out the upper bed of coal on either side of a *gate road* in short faces of say three or four yards, leaving *stoops* to protect the roof and roads.

BACK-STAY (Y.). A wrought-iron forked bar attached to the back of trams when ascending an inclined plane, for throwing the trams off the rails in the event of a rope or coupling giving way. See Fig. 4.

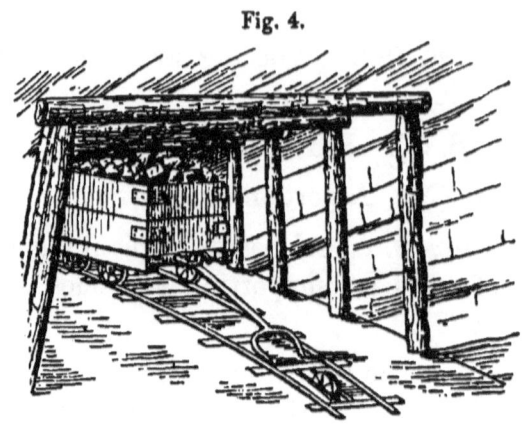

Fig. 4.

BAFF-ENDS. Long wooden wedges for adjusting *tubbing plates* or *cribs* in *sinking pits* during the operation of fixing the *tubbing*.

BAFFLE (M.). To brush out or mix fire-damp with air in order to render it non-explosive; a dangerous practice, and not now allowed.

BAFFLER (N. S.). The lever with which the throttle-valve of a *winding engine* is worked.

BAFF-WEEK (N.). The week next after the pay week, if wages are paid fortnightly.

BAG (S. S.). A quantity of *fire-damp* suddenly given off from the coal.

BAG COAL. Coal put into coarse canvas bags and sold in small quantities.

BAG OF FOULNESS (N.). A cavity in a *coal seam* filled with *fire-damp* under a high pressure, which, when cut into, is given off with much force, and danger of causing an *explosion*.

BAILIFF. Name formerly used for *manager* of a mine.

BAIT (N.). Food taken by a collier during his *shift*.

BAIT-POKE (N.). A bag for *bait*.

BALANCE. The counterpoise or weights attached to the drum of a *winding engine*, to assist the engine in lifting the load out of the *pit bottom*, and in helping it to slacken speed when the *cage* reaches the surface. It consists often of a bunch of heavy chains suspended in a shallow *shaft*, the chains resting upon the pit bottom as unwound off the balance-drum attached to the main shaft of the engine.

Fig. 5.

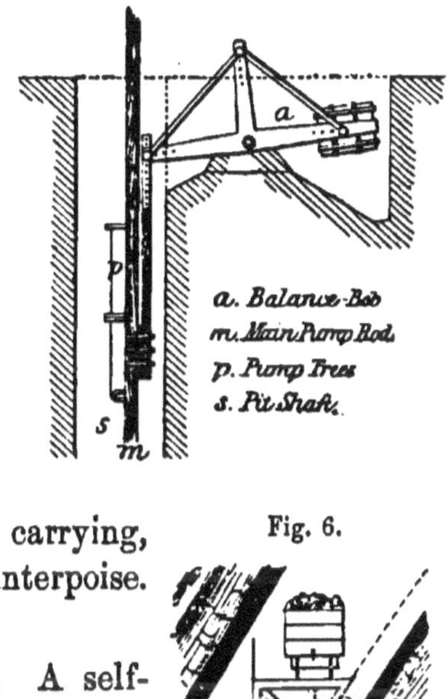

*a.* Balance-Bob
*m.* Main Pump Rod
*p.* Pump Tree
*s.* Pit Shaft.

BALANCE-BOB. A large beam or lever attached to the main rods of a *Cornish pumping engine*, carrying, on its outer end, a counterpoise. See Fig. 5, *a*.

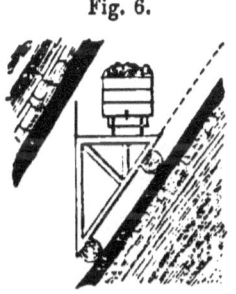

Fig. 6.

BALANCE-BROW (N. S.). A self-acting inclined plane in steep seams, which is driven on the full rise of the mine, and down which the tubs of coal are lowered and the empties elevated upon a kind of carriage or platform on wheels actuated by a rope or chain from above. See Fig. 6.

BALANCE-PIT. The pit or shaft in which a *balance* rises and falls.

BALK. 1. A more or less sudden thinning out of a seam of coal, not unfrequently 100 yards in width. See diagram, Fig. 7.

Fig. 7.

2. A bar of timber for supporting the *roof* of the mine, or for carrying any heavy load.

BALL IRONSTONE (S. S.). Strata containing argillaceous *ironstone* in the form of nodules, which range in weight up to 15 or 20 cwt.

BALLSTONES (N. S.) Ancient term for *ironstone*.

BALNSTONE (N.). Stone or rock forming the *roof*.

BAND (S. S.). 1. A *winding rope* or chain.

2. A seam or thin stratum of stone, &c., often interstratified with coal.

3. (C.) A bed or seam of coal.

BANDFUL (S. S.). A cage or strictly speaking a rope load, e. g. a *bandful of men*, by colliers commonly pronounced *bontle*.

BANDSMAN (S. S.). A loader or filler of coal, &c., underground.

BANGING-PIECES. See *Catches*.

BANK. 1. The top of the pit, or out of the pit.

2. The surface around the mouth of a shaft.

3. To manipulate coals, &c., on the bank.

4. The whole or sometimes only one side or one end of a *stall* or *working place* underground.

5. (C.) A large heap or stack of mineral on surface.

USED IN COAL MINING, ETC. 11

BANK-HEAD. The upper end of an inclined plane next to the engine or *drum* (2), made nearly level. See Fig. 8.

Fig. 8.

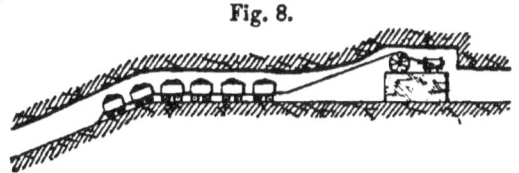

BANK-HOOK (M.). An iron hook with which the *banksman* pulls the full *tubs* off the *cages*.

BANKING. 1. (M.) Sorting and loading of coals at *bank* (2).

2. (C.) Heaping up minerals on surface for future sale.

BANK LEVEL (Y.). The level *heading* out of which *banks* (4) are worked.

BANK OUT (N.). To stack or stock coals at surface when short of wagons, &c., to load into.

BANK PLATES. Cast-iron sheets with which a *heapstead* or pit bank is laid or floored for the more expeditious manipulation of the *tubs*.

BANK-WORK (Y.). A system of working coal in South Yorkshire (shown in plan in Fig. 9).

Fig. 9.

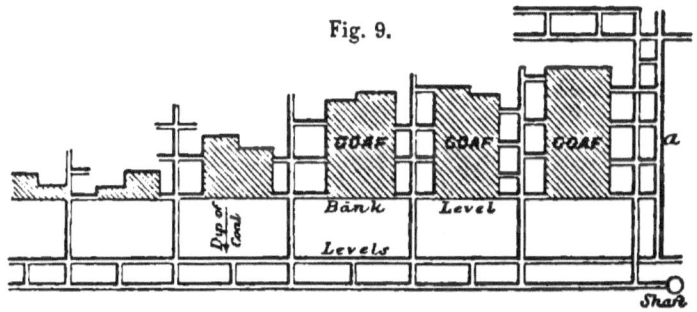

BANKSMAN. The man in attendance at the pit top for superintending the work of *banking*.

BANKSWOMAN (S. & N. W., S., L.). A female employed at *bank* (1) to pick the stones from and to clean the coals for the market.

BANK TO BANK. A period occupied by a *collier* between leaving the *bank* (1) and returning to same. A *shift*.

BANNOCK (Sh.). Brownish grey clay suitable for making into firebricks.

BANNOCK (S. S.). To *hole* on the top of a seam.

BANT (D.). A certain number of men, usually three or four, who in former times, prior to the introduction of *cages* and *conductors*, used to ride up and down in a *pit-shaft*, sitting in short loose pieces of chain attached to a hemp rope in a cluster, with their knees pointing inwards toward the centre of the *shaft*. There were usually two *bants*, the lower or *bottom bant* which was composed of men, and the upper or *foaley bant* which was made up of a cluster of lads fastened a few feet above the heads of the men. There was only one rope used for raising and lowering men; the second was a chain, which was sent *up empty*, or without anything attached to it, when men were descending, and *vice versâ*.

When the *bant* was used, at some collieries the *winding-ropes* or rather chains were pulled close up to the sides of the *shaft*, and the man-rope *drum* (1) was put in gear, the *bant* working over a third *pulley* in the *pit frame*. See *Hold out!* and *Tucklers*.

BAR. A length of timber placed horizontally for supporting the *roof*. In some cases *bars* of wrought iron, about 3" × 1" × 5', are used.

BARE. To strip or cut by the side of a *fault*, boundary *hollows*, &c.

BARFE SATURDAY (N.) The word *barfe* = off. The Saturday upon which wages are not paid.

BARGAIN-WORK (N.). Underground work done by contract, e. g. *heading*, road laying, &c.

BARING. 1. The surface soil and useless strata overlying a seam of coal, clay, ironstone, &c., which is being worked by *open-hole*, which has to be removed or *bared* preparatory to working the mineral.

2. (Y.) *Holing*, which see.

3. (Y.) Using a stout iron bar to get the Cleveland ironstone down, after blasting.

BARITELS (F.). See *Horse-gin*.

BAROMETER HOLIDAY (D.). Any day on which, owing to the very low state of the barometer (for instance, when it sinks below say 29 inches), much *fire-damp* may naturally be expected to be given off in the mine, causing risk of *explosion*, no work is carried on underground.

BARREN GROUND. Strata unproductive of seams of coal, &c., of a *workable* thickness.

BARRIER. A solid block or rib of coal, &c., left unworked between two collieries or mines for security against accidents arising from the influx of water from one to another; in width often as much as 100 yards.

BARRIER SYSTEM (N.). The most modern and approved method of working a colliery by *pillar and stall*, where solid ribs or *barriers* of coal are left in between a set or series of *working places*; the width of such barriers being from 40 to 50 yards. See plan, Fig. 10.

BARRING. 1. The timbers in the workings for keeping up the *roof*.

2. (S.) The timber walling or casing of *pit-shafts*.

BARROW-MAN. One who, in former times, used to convey coals underground in a wheelbarrow from the working places to the *rolley-ways*.

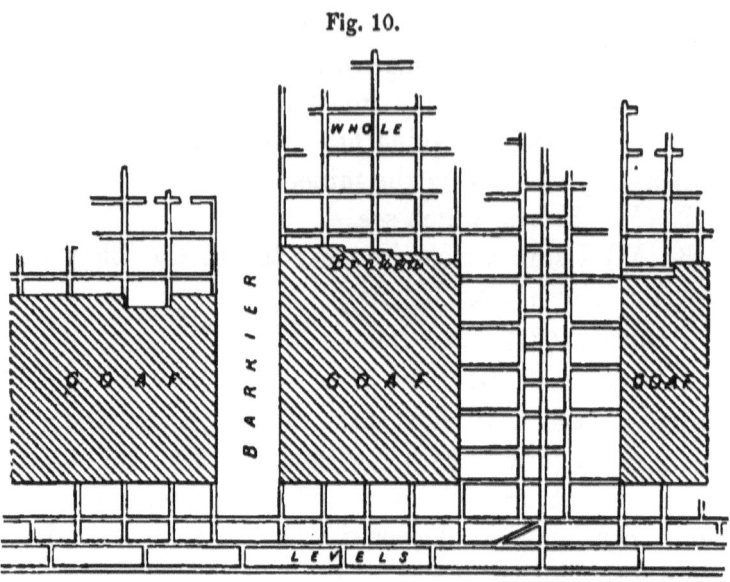

Fig. 10.

BARROW-WAY (N.). The underground roads along which the *barrow-men* worked.

BASH (S. W.). To fill with rubbish the spaces from which the coal has been worked away.

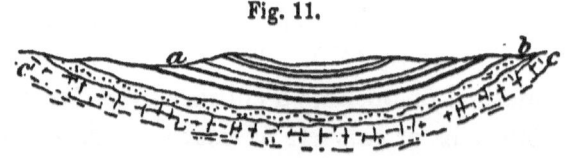

Fig. 11.

*a*, Coal Measures.   *b*, Millstone Grit.   *c*, Carbonaceous Limestone.

BASIN. A coal-field having some resemblance in form to that of a basin. The Forest of Dean coal-field is perhaps the most perfectly basin-shaped one in Great Britain. See diagram, Fig. 11.

BASKET (L.). A measure of weight = 2 cwt. occasionally used in East Lancashire.

BASKETS (S. S.). Shallow pans into which *small* is raked by *fillers* for loading into *tubs*.

BASS. Black carbonaceous shale.

BASSET. 1. *Outcrop*, which see.
2. Shallow or *rise* side of a working.

BASSET-EDGE. The actual outcrop or boundary of a seam, where it appears at the surface.

BAT (L., S. S.). See *Baffle*. *Batting out gas* was formerly a regular though unsafe thing to do.

BATE (S. S.). To excavate or cut away the *floor* of a mine.

BATE BARREL (Lei.). After drawing a number of barrels of water out of a *sump*, the first barrel that there is not sufficient water to fill is called the *bate barrel*.

BATE-WORK (N.). Short work.

BATT. See *Bass*.

BAUM-POTS (Y.). Calcareous nodules found in the shale forming the roof of the "Halifax Hard" coal seam.

BAY. 1. An open space for a *gobbin* or *waste* between two *packs* in a *long-wall* working.
2. (L.) A *board*, which see.

BAYSHON (Som.). An air *stopping*, which see.

BEANS (N.). All coal which will pass say a half-inch screen or mesh.

BEARERS (S.). Women formerly employed to bear or carry coals out of the mines upon their backs in *creels*, for which they were paid from 1*s*. to 1*s*. 2*d*. per day, finding their own *creels* and candles.

BEARING DOOR. A door placed for the purpose of directing and regulating the amount of ventilation passing through an entire district of the mine.

BEARING IN (S.). The depth or distance under, of the *holing* or *kirving*.

BEARING-UP PULLEY. A pulley wheel fixed in a frame and arranged to tighten up or take up the slack rope in *endless rope* haulage.

BEARING SYSTEM. The employment in former times of females to carry out upon their backs the produce of the mine.

BEARS (D.). Calcareous clay-ironstone in nodules.

BEATER. 1. (N.) An iron rod for *stemming* the hole preparatory to firing a *shot*.

2. (M.) A wooden mallet for consolidating, or making air-tight, the clay, when building *wax* walls or *dams*.

Fig. 12.

BECHE or BITCH (N. E.). A hollow conical-headed iron rod for extricating boring rods from *bore holes* (1). See Fig. 12.

BED. 1. The level surface of rock upon which a *curb* or *crib* is laid.
2. A stratum of coal, ironstone, clay, &c.
BELL. 1. To signal by ringing a bell.
2. (F. D.) See *Bell-mould*.
BELLED. The widened out portion of a pit shaft at the *inset* in order to give plenty of room for running the trams past the shaft, and for changing them in the *cages*.

BELL-MOULDS, BELL-MOUTHS (Som.). Conical-shaped patches of the *roof*, being probably the bases of the fossils called *sigillaria*, or the roots of trees.

BELL-PIT (D.). Pits working argillaceous ironstone by the system called *Bell-work*, which see.

BELL-SCREW or SCREW BELL. An internally threaded bell-shaped iron bar, for recovering broken or lost rods, &c, in a deep *bore hole* (1). See *Beche*.

BELL WORK (D.). A system of working ironstone *rake* measures by underground excavations, around the pits or shafts in the form of a *bell* or cone. Pits are sunk about 20 to 40 yards apart, the ironstone is then worked away between the pits and lastly taken from the sides of the shafts, thus forming them into *bells*. See diagram, Fig. 13.

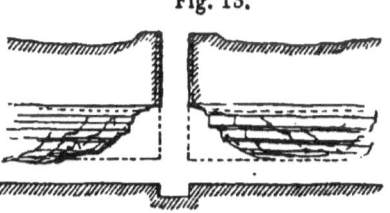

Fig. 13.

BENCH (Pa.). 1. A small *tram* or *car* of about 7 cubic feet capacity used in the *breasts* for carrying coal from the *face* of the *workings* to the shoot or *chute* down which it is *dumped* to the *gangway* platform for reloading into larger *cars*.

2. (Lei.) To wedge the *bottoms* up below the *holing*.

3. A stratum of coal forming portion of a *seam*; some *seams* are made up of a number of *benches* separated by strata of shale, &c.

BENCHERS (S.). Men who are employed at the bottom of inclined planes in the mine.

BENCHING. 1. See *Holing*. Also to break up with wedges the bottom coals when the holing is done in the middle of the seam. See Fig. 14.

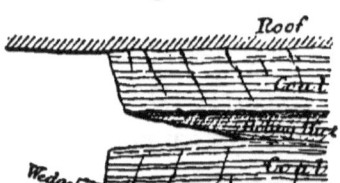

Fig. 14.

2. (Ch.) The lower portion of the rock-salt bed worked in one operation (up to 12 feet in thickness).

BENCH WORKING. The system of working one or more seams or beds of mineral by *open working* in stages or steps as shown in diagram, Fig. 15.

Fig. 15.

"BEND AWAY" or "AWAY!" (N.) Raise the cage in the *shaft*.

BENK (D.). See *Bank* (4).

BENT (S.). Subsidence of *roof* having taken place to rear of *working face*, e.g. a *bent roof*.

BERGMEISTER (Pr.). An *Inspector* of mines.

BETRIEBSFUHRER (Pr.). The mining engineer or *Manager* of a coal mine, who is personally responsible for the safety of the *workings*. He sometimes acts as an *Obersteiger*.

BETRIEBSPLAN (Pr.). A sketch or rough *plan* of underground *workings*, proposed to be executed during the next 12 months, submitted for approval to the *Revierbeamt*.

BIAT or BYAT. A timber stay or beam in a *pit shaft*.

BIBBLEY ROCK (S.S.). Conglomerate or pebbly rock.

BIGGIN (N.). A built-up pillar of stone or other *débris* in a working place or heading for a support to the roof, e.g. *bigging the gob* means, building a *pack* in a worked-out place in a pit.

BILL DAY (N.). That on which *viewers* examine the colliery accounts, &c.

BILLET (Som.). A short prop or *tree* of timber.

BILLY. 1. (F. D.) A box for holding ironstone, carried by a boy in the mine.

2. See *Billy Playfair*.

BILLY BOY (S. W.). A lad who attends to the working of a *Billy Playfair*.

BILLY PLAYFAIR or FAIR-PLAY (S. W.). A man's name given to a mechanical contrivance for weighing coal, consisting of an iron trough with a sort of hopper bottom, into which all the *small* passing through the *screen* is conducted and weighed off and emptied from time to time.

BINCHING. 1. (Som.) The stone upon which a vein of coal rests.

2. See *Benching*, also *Undercutting*.

BIND or BINDS. 1. Indurated argillaceous shale or clay, very commonly forming the roof of a coal seam and frequently containing clay ironstone.

2. (N.) To hire.

BINDER. See *Bind* (1).

BINDING (N.). Hiring of men for pit work.

BING. 1. (S.) A place where coals, &c., are stocked, or *débris* tipped at surface.

2. (S.) To put coals on one side in wagons or in stacks at surface.

BIT. A piece of steel placed in the cutting edge of a drill.

BITUMINOUS COAL. A clear and free-burning variety of coal, or a flaming coal of a fuliginous character.

BLACKBAND. Carbonaceous *Ironstone* in beds, mingled with coaly matter sufficient for its own calcination.

BLACK-BATT. Black carbonaceous shale.

BLACK COTTON (In.). Soil from 6 to 10 feet in thickness overlying the *coal measures*, which in dry weather opens and cracks up like fissures.

BLACK-DAMP. Carbonic acid gas, much the same as *after-damp*. It will not support combustion, and is very deadly.

BLACK DIAMONDS. A term frequently applied to signify *coal*.

BLACK-JACK (D.). A kind of cannel coal.

BLACK MUCK or BLACK MOULD (L.). A dark-brown powdery substance, consisting of silica, alumina, and iron; found in iron mines.

BLACK-RING (S. S.). In a *sinking-pit*, it means a thin bed or shed of coal as seen running round the shaft sides, having the appearance of a black circle or ring.

BLACKS (Som.). Soft dark-coloured shale.

BLACKSTONE (N.). Highly carbonaceous shale.

BLAST. 1. The sudden rush of fire and gas and dust of an *explosion* through the underground workings and roadways of a colliery.

2. To cut or bring down coal, rocks, &c., by the explosion of gunpowder, dynamite, &c.

BLAES or BLAIZE (S.). A hard-bedded sandstone, free from joints; also a kind of *underclay* with balls of ironstone; also ordinary *bind*.

BLECK (N.). Pitch or tar upon ropes.

BLEED. A coal or other stratum is said to *bleed* when it gives off water or gas.

BLIND. 1. (F. D.) See *After-damp*.

2. (S.) To erect a *stopping* in a *bolt-hole* or other underground roadway.

BLIND COAL. Coal altered by the heat of a *trap dyke* into something resembling *anthracite*.

BLIND-PIT (L.). See *Drop-staple*.

BLIND-ROAD or BLIND-WAY (M.). Any underground roadway not in use either for drawing coals, &c., ventilation, or for travelling along, having *stoppings* placed across it.

BLOCK COAL. Coal in large lumps.

BLOCKY (B.). See *Block Coal*.

BLOW. 1. To blast with gunpowder, &c.

2. A *dam* or *stopping* is said to *blow* when gas escapes through it.

3. (Y.) A *roof* is said to *blow* when it commences to break in or *weight*.

BLOWER. 1. A sudden emission or outburst of *fire-damp* in a mine, the gas generally coming out of the coal. They frequently continue to *blow* (2) for many days or weeks. The pressure of the gas is at first not unfrequently as high as 300 or 400 lb. per sq. in., but gradually decreases. The quantity of gas given off is sometimes of enormous volume, filling a great portion of the workings of an extensive colliery in a few seconds only, and extinguishing nearly every lamp in the mine.

2. A man who blasts or fires shots in a pit, or who drills the holes and charges them, ready for firing.

BLOW-GEORGE. A small centrifugal fan worked by hand, for airing or ventilating a *heading* or *pit*.

BLOWING ROAD (S. S.). *Intake* or fresh-air road in a mine.

BLOWN-OUT SHOT. In blasting, when it occurs that the coal or rock bears the strain of the ignited explosive longer than the *stemming* in the hole, the result is called a *blown-out shot*, or one that has gone off but not done its work.

BLOWS (L.). Frequent and sudden risings of quicksand in sinking through watery ground.

BLOW-UP. 1. An *explosion* of *fire-damp* in a mine.

2. To allow atmospheric air to get access to certain places in coal mines, so as to generate heat, and ultimately to cause *gob fires*. This is to *blow up* a *fire* (4).

BLUE BIND. See *Bind* (1).

BLUE CAP. The blue or brownish-coloured halo of ignited gas (fire-damp and air) on the top of the flame of a safety lamp. To carry on work in an atmosphere which shows a *cap* is unsafe.

BLUE GROUND (S. S.). Strata of the *coal measures*, consisting principally of beds of *bind* (1).

BLUE METAL (N.). See *Bind* (1).

BLUFT (Lei.). To extinguish or put out of sight a candle or other light.

BLUE STONE (S. W.). In Caermarthenshire it is a name for *bind* (1).

BOARD or BORD. 1. (N.) A wide *heading*, usually from 3 to 5 yards.

2. (Y.) When a seam of coal is worked parallel to the natural joints or *faces* intersecting it, it is said to be worked *board*.

3. A plane of cleavage in coal, the line of which is generally more or less north and south.

4. A piece of board with the word *Fire* or *Danger*, or some other notice in reference to *gas, safety lamps, shot-firing*, dangerous *roof,* &c., painted upon it, to warn the men and boys in the *workings*. It is hung by a nail to a *prop*, or fixed in some other conspicuous position, beyond or behind which the danger lurks.

BOARD AND PILLAR. A system of working coal where the first stage of excavation is accomplished with the roof sustained by coal. The coal is worked out to the extent of from say 30 to 60 per cent. of the whole seam. Of course, this system is capable of very great modification, and the size of pillars is determined by the circumstances under which the system is carried out. Fig. 16 is a sketch plan, showing an arrangement of the workings.

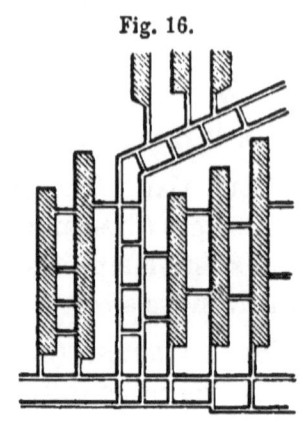

Fig. 16.

BOARD COAL. Coal having a fibrous or woody appearance. Of the Secondary and Tertiary eras.

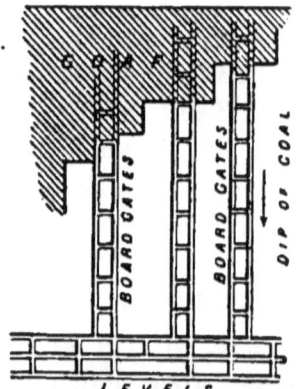

Fig. 17.

BOARD GATES (Y.). *Headings* driven in pairs generally to the *rise*, out of which *banks* (4) or *stalls* are opened and worked. See plan, Fig. 17.

BOARD AND WALL. See *Board and Pillar*.

BOARD-ROOM (S.). A *heading* driven *board* (2).

BOARD-WAY'S COURSE (N.). At right angles to the planes of cleavage of the coal. See *Face on*.

BOAT COAL (Pa.). Coal which is loaded into boats on canals, rivers, &c.

Bob. An oscillating bell-crank or lever, through which the motion of an engine is transmitted to the pump-rods in an engine or pumping-pit. (See elevation of ⌊ 'bob,' fig. 18. There are ⊥ bobs, ⌊ bobs, and ∨ bobs.

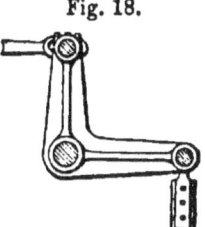

Fig. 18.

Bogie. 1. (Y.) A small truck or trolly upon which a *kibble* is carried from a *sinking pit* top to the *spoil bank*.
2. A weighted truck run foremost or next to the rope in a *set* or *train*.

Boll (N.). An ancient measure for coal, containing 9676·8 cubic inches, or $\frac{1}{440}$ part of a *Ten*.

Bolt or Bolt-hole (S. S.). A short narrow *heading*, connecting two others.

Bond. 1. (N.) Agreement for hiring workmen.
2. (F. D.) A *wind* (5) made by a *winding engine*.
3. (N. S.) A bed, band, or seam of ironstone.

Bone (Pa.). Hard slaty carbonaceous beds of rock.

Bonnet. 1. The overhead cover of a *cage* or *swinging bont* usually constructed in the form of a ridge tile ∧ so as to ward off the blows from anything accidentally falling down the *shaft*.
2. (S.) See *Bell-mould*.

Bonnet Roller, Bonnet Pulley, Bonnet Sheaf. See *Hat Roller*.

Bont or Bond. The cage and winding rope with attachments.

Bontle (M.). A cage-full of men.

Boobey (Som.). A kind of box holding 6 to 8 cwt. of coal in which dirt or rubbish is sent to *bank* (1).

BOOLIES (N.). A collier's term for brothers.

BOOT LEG (L.). A short pipe of leather through which the water is drawn from a *pot-hole* into a pump of a *sinking set* (1).

BORD (Y.). A road or heading in a pit in *board and pillar* workings.

BORDS AND LONGWORK (Y.). A system of working coal in the manner shown in Fig. 17. The *modus operandi* is briefly as follows:—

Firstly, the main levels are started on both sides of the shafts and carried towards the boundary.

Secondly, the *boardgates* are set away in pairs to the *rise* and continued as far as the boundary, or to within a short distance of a range of upper levels and other *boardgates*.

Lastly, the whole of the *pillars* and remaining coal are worked out downhill to within a few yards of the levels, and ultimately the coal between the levels is worked away.

BORE. 1. To prove, by boring vertical holes, the character and thickness of strata.

2. The proportion of the sectional area of a pipe filled with running water. When a pipe is discharging water to its greatest capacity, i.e. when the pipe is quite full, it is said to be *running full bore*.

3. A *Borehole* (1), (2), (3).

BORE-HOLE. 1. A hole made with a drill, auger or other tools, from 1 in. to as much as 30 ins. diameter, and to a depth of several thousand feet (5500 feet having been attained at Potsdam in America), for exploring strata in search of minerals, for water supply, and other purposes.

2. A hole bored into the *face* of a coal wall or stone *drift*, &c., for blasting purposes.

3. Holes bored in *ribs* and *pillars* for proving the position of old workings, proving *faults*, letting off accumulations of gas or of water.

BORE MEAL. Mud or finely chopped-up *débris* out of a *bore-hole*.

BORING-HEAD. The group of chisels or cutters by which the strata are cut through in boring. See *Bore* (1), Fig. 19.

BORING RODS. Square iron rods of Swedish iron of the toughest quality, made in lengths of 4 or 5 yards, having male and female screws at the extremities for connecting them together in a bore-hole. See Fig. 20.

BOSH (Water Bosh) (S.W.). A tank or tub out of which horses drink.

Fig. 19.

Fig. 20.

BOTTLE-JACK. An appliance for raising heavy weights in a pit.

BOTTOM. The bottom of the shafts and roadways, &c., near the shafts.

BOTTOMER. The person who loads the cages at the pit bottom, and gives the signals *to bank* (1).

BOTTOM PILLARS. Large blocks of solid coal or *mine* (1), left unworked round about the *pit shaft*. See *Shaft Pillar*.

BOTTOM STEWARDS (Y.). Underground officials.

BOTTOMS (M.). The lowermost portion or natural division of a *seam* of coal, &c. The holing is sometimes done above the *bottoms*, and then they are *benched* (2), up.

BOULEUR (Belg.). Small girls who collect the coals into heaps in the working places underground to be filled into *trams* by older girls.

BOUTONS (S.). Masses of roof stone or shale.

BOUT. 1. (M.) A coil of rope upon a *drum*.

2. (Lei.) A dinner or other jollification given by the owners or lessees of a colliery to their colliers and other workmen in honour of some special event, e.g. finding of coal, a coming-of-age, &c.

BOW. The bent iron bar or handle suspending the body of a *kibble*.

BOWK. An iron barrel or tub in which the *débris* from a *sinking pit* is raised. See Fig. 21. It is attached to the rope by three short chains with hooks, and holds about half a ton of stuff.

Fig. 21.

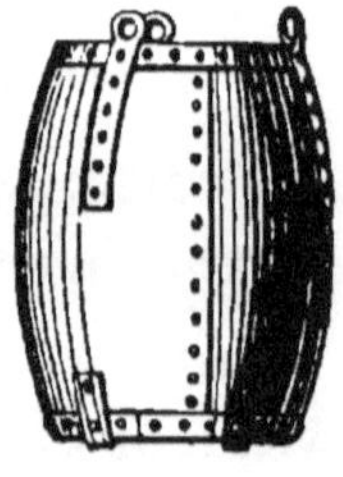

BOX. The vehicle in which coals, &c., are conveyed from the *working places* along the underground roadways up the *shaft* and to the unloading places at *bank* (1).

It has a capacity of from 8 to 20 cwt., varying according to the thickness of the seam worked, and the height and width of the roads; and weighs from 3 to

6 cwt. The wheels are from 10 to 15 inches in diameter and made of cast steel, the framework and bodies are of ash and elm strengthened with iron ribs and plates.

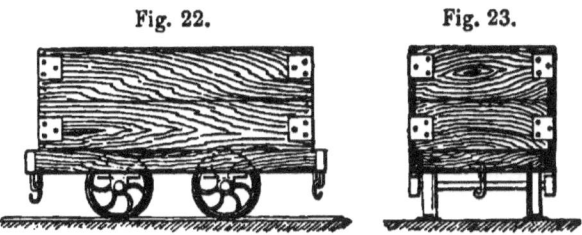

Fig. 22.    Fig. 23.

Figs. 22 and 23 show a side and an end elevation of a box as commonly constructed.

BOX BELL. See *Bell-screw*.

BOX BOTTOMS (Lei.). The small coal or slack which falls to the bottom of the *boxes* or *tubs*. It is produced by breakage in transit underground, and by sorting on the *bank* (1).

BOXED OFF. Enclosed or protected by a wooden pipe or partition.

BOXES (Pa.). Wooden partitions for conducting the ventilation from place to place.

Fig. 24.

BRACEHEAD. Wooden handles or bars for raising and rotating the rods when boring deep holes. (See Fig. 24.) The handles are firmly set in an iron socket, forming the uppermost end of the top rod, a short chain being attached to the ring on the top by which the rods are suspended from the *brake staff*. Sometimes four handles are employed set cross-ways.

BRAKE. 1. A stout wooden lever to which boring rods are attached, and is worked by one or more men.

2. (N. S.) To lower *trams* down *dips* (4) by means of a wheel and rope.

BRAKESMAN (N.). The man who works the *winding engine*.

BRAKE-STAFF. See *Brake* (1). It has an up-and-down motion, imparted to it either by machinery or by hand.

BRAKING (N.). Working a *winding engine*.

BRANCH. 1. (Som.) An underground road or heading driven in *measures*. See diagram, Fig. 25.

2. A roadway underground branched off from a *level*, &c.

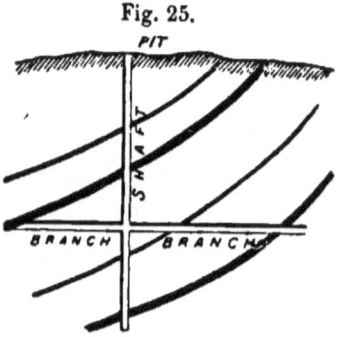

Fig. 25.

BRASHY. Short and tender, as *brashy bind*, &c.

BRASS. Iron pyrites in coal. Occurs generally in lenticular patches, small veins, and scaley partings.

BRAT (N.). A thin bed or band of coal mixed with lime and iron pyrites.

BRATTICE. 1. A division or partition in a *shaft*, *heading*, or other underground *working place*, for providing for *ventilation*, &c. It divides the place into two parts, one for the ingress of the fresh air, and one for the egress of the vitiated air. A *brattice* may be constructed of brick or stone work, of coarse clothing nailed to timbers, or of sheet-iron tubes about 18 inches in diameter, or of boarding. Figs. 26 and 27 show

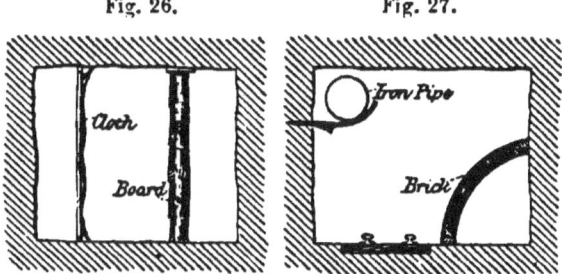

Fig. 26. Fig. 27.

cross sections of four ways of making a *brattice* in a *heading*. Strictly speaking the iron pipe system is not a *brattice*.

2. (M.) A built-up pillar of cordwood something like a large *chock* (which see), and serving a similar purpose. Called also *brettice* and *brittice*.

BRATTICE-ROAD. A *gateroad* through the *goaf* supported by *brattices* (2) or timber *packs*. Fig. 28 gives a cross section of a roadway of this description.

Fig. 28.

BRATTICE WALL. The *bratticed* side of an *aircourse* or other road.

BRAZZIL (M.). See *Brass*.

BREAK. 1. A crack or small natural cavity or fracture in the coal seam.

2. A crack, often several inches in width, proceeding from old workings or *hollows*.

BREAK IN (S.). To commence to *hole*.

BREAKAGE CLAUSE. A clause inserted in some mining leases providing for an abatement of royalty or allowance on weight for a certain weight of small coal or *breakage* sent out in every ton of large coal, e.g. 120 lbs. in every 2640 lbs. or collier's ton.

BREAKER. 1. (N.) A large crack formed in the roof next to the *goaf*. See *Break* (1).
2. (Som.) A coal getter or "hewer."
3. (I.) A collier who wedges down coal and fills it into *tubs*.

BREAKER BOY (Pa.). A lad who attends to a coal-breaking machine.

BREAKING BAND (S.). A method of *setting* or fixing props in the workings, in lines running diagonally to the line of the *face* or *wall*.

BREAKING-DOWN MACHINES. Mechanical appliances, such as wedges, &c., worked by compressed air or by hydraulic power, for bringing down the coals after they are *holed*.

BREAKING UP (Cl.). A system under which a skilled miner engages an unskilled man, the former paying the latter a mere labourer's wages until he becomes able to demand the wage that experience has made him worth.

BREAK OFF. To drive a *thirl* or *bolt-hole*, &c., out of a *gate-road, level,* &c.

BREAK UP (M.). To cut away and remove the *floor*.

BREAST. 1. (Pa.) A stall 10 yards in width.
2. (I.) A *stall* in a *steep seam* from 12 to 18 yards wide. They are carried one above another from the lowest level to the *rise*. Fig. 29 shows a section of three *Breasts* with the unworked coal between them.

Fig. 29.

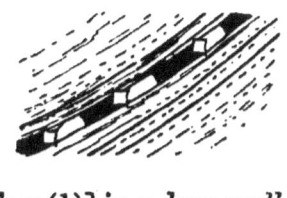

3. (Lei.) To take down or *get* a *buttock* of coal *end on* [i.e. working it off in a direction at a right angle with the line of the *Face* (1)] in a *long-wall stall* when the *roof* has fallen in close up to the working face, thus preventing work going on in the ordinary way.

BREAST AND PILLAR (Pa.). A system of working *anthracite* coal by *boards* 10 yards in width, with narrow pillars 5 yards wide between them, *holed* through at certain intervals. See *Board and Pillar.* The *breasts* are worked from the *dip* to the *rise*.

BREAST-BORE (S.). A *borehole* (3) put in parallel with the *seam*, made and kept in advance of a working-place, for the purpose of ascertaining the position of old works, tapping water, letting off gas, &c.

BREAST-EYES (L.). See *Day, Day-hole.*

BREAST-HEADS. Natural joints in rock, coal, &c.

BREASTING. 1. (N. S.) A short leading *stall*, worked at right angles to, and forming the *face* (1), of the main levels.
2. Wide *heading* or *level*.

BREATHER. An apparatus brought out by a Mr. Fleuss for use in impure atmospheres, enabling a man

D

to enter and explore underground workings filled with noxious gases. It consists of a mask or mouthpiece, a knapsack, and an elastic air-reservoir or bag, and is charged with oxygen gas, which the wearer inhales, and, by an ingenious arrangement, breathes over and over again; and consequently can remain in gas for several hours at a time (Fig. 30). A special form of *safety-lamp* is used with the *breather*, constructed upon the same principle.

Fig. 30.

BREECHING (M.). Drawing loaded *trams* down hill underground.

BREEDING FIRES (S. S.). Spontaneous combustion in a mine. See *Gob Fire*.

BREESE (S.). Fine slack.

BRICK COAL. Small and rough quality of coal suitable for brick kilns and similar purposes.

BRICK FUEL (S. W.). Patent Fuel.

BRICKING. The *walling* or casing of a *pit-shaft*.

BRIDAL (S.). A contrivance for preventing *tubs* from overturning upon steep inclined planes (1 in 3 or 4).

BRIDGE. 1. See *Air Crossing*.

2. A platform on wheels running upon rails, for covering the mouth of a *pit-shaft* when landing coal, *débris*, or men at surface.

BRIDLE CHAINS. Short chains by which a *cage* is attached to a *winding rope*. Either four or six are used.

BRIERS (N.). Beams or girders fixed across a shaft top.

BRIGHT-HEADS (Y.). *Backs* (1) or *slines*.

BRING-BACK. To work away the pillars of coal or the *broken* from the boundary towards the pit bottom.

BRIQUETTES (Belg.). See *Brick Fuel*.

BRITCHING (S.). Horse's tackle used when the *tub* precedes the horse upon a steep incline.

BRITISH (S.). A kind of *pack* or *building*.

BROADSTONE BIND, &c. *Bind* (1) which breaks up into large blocks or slabs.

BROADWALL (N.). See *Longwall*.

BROBS (M.). Short thick timber props or *sprags* for supporting the coal whilst it is being *holed*. They are set about half way under the *holing*. See Fig. 31.

Fig. 31.

BROKEN. That part of a mine where the mineral has already been partially worked away, and where the remainder is in course of being extracted. See Fig. 10.

BROKEN GROUND. Faulty or unproductive *measures*.

BROKEN JUD (N.). A *jud* in course of being worked off from the *whole*.

BROW. 1. (L.) An underground roadway leading to a working-place, driven either to the *rise* or to the *dip*.

2. A low place in the roof of the mine, giving insufficient *head-room*.

BROW-BAR (M.). A massive *curb* or beam of timber fixed in the *walling* of the shaft across the top of the *inset*.

BROWN COAL. Woody or peaty-looking coal of a brown or black colour found in the Secondary and Tertiary rocks.

BROW UP (L.). An inclined roadway driven to the *rise*. See *Brow* (1) and *Upbrow*.

BRUSH. 1. (M.) To mix *gas* with air in the mine by buffetting it with a jacket, &c. This is done to render it inexplosive. It is a very dangerous practice, and not now allowed.

2. (F. D.) A rich brown hæmatite iron ore.

3. (Som.) See *Altogether Coal*.

4. (S.) To take down or *rip* the *roof*.

BRUSHERS (S.). Men who *brush* (4) the *roof*, build *packs* and *stoppings*, which work is called *brushing*.

BRUSHING-BED (S.). The stratum *brushed* or *ripped*.

BRUSKINS (M.). Small coal in lumps about a pound in weight each.

BUCKET. The top valve or clack of a lifting *set* (1) of pumps. It is attached to the lower end of the rods, and works within a long pipe or barrel. Fig. 32 is a plan and side view of an ordinary pump *bucket*.

Fig. 32.

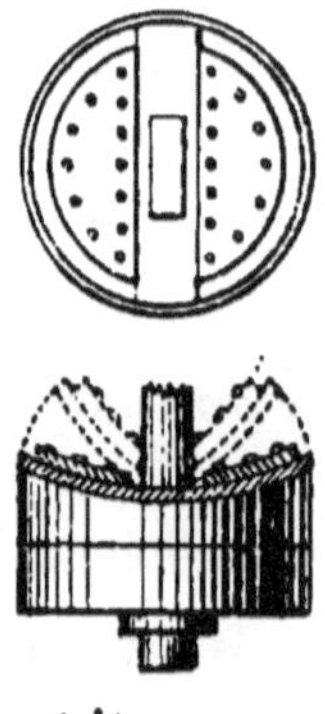

BUCKETING. The operation of taking out a worn-out pump bucket or *clack*, and replacing it with a new one, in connection with pumps fixed in an *engine-pit*, or belonging to the Cornish system of pumping.

BUCKET SWORD. A wrought-iron rod to which a pump *bucket* is attached, having at its upper end a *knocking-off joint*.

BUCKET-TREE. The pipe between the *working barrel* and the *windbore*.

BUCK WHEAT (Pa.). Anthracite which will pass a *screen* varying in width between ⅞ and ¼ of an inch.

BUGGIED (Pa). *Trammed* or *put*, which see.

BUGGY (Pa.). A small *car* or *tram* of about 7 cubic feet capacity, used in the *breasts* for conveying the coal from the *faces* to a shoot, or *chute*, down which it is *dumped* to the *gangway* platform for reloading into larger *cars*.

BUILDERS-UP. Men who make *packs*, set timber, &c., in some ironstone mines.

BUILDING (S.). A built up block, or pillar of stone or coal to carry the *roof*.

BUILDING-STONE (S.). Sandstone or *bind* (1) suitable for *pack building*.

BULK. 1. (B.) See *Dip*.
2. Coal in large and small lumps in large quantities.

BULKHEADS. See *Chock*.

BULL (N.). 1. An iron rod for preparing a *shot*-hole in watery ground, and when the hole has to be lined with clay. Using a *bull* is called *bulling*.
2. See *Backstays*.

BULL ENGINE. A single-acting pumping engine constructed upon the direct-acting principle, that is to say, it has no beam or toothed gearing, the cylinder being inverted and fixed directly over the *pit-shaft*, the pump-rods forming a continuation of the piston-rod.

BULLER SHOT (S.). A second one put in close to and to do the work not done by a *blown-out shot*, loose powder being used.

BULLIONS (L.). Nodules of clay ironstone, iron pyrites, shales, &c., which generally enclose a fossil.

BULL-WHEEL (Pa.). A wheel upon which the rope carrying the boring rods is coiled when boring by steam machinery.

BUMP. A very sudden breaking, sometimes accompanied by a settling down, or upheaval of, the strata, during the working away of the mineral, accompanied by a loud report or bumping noise heard in the mine.

BUMPERS (M.). See *Catches* (3).

BUNKERS (S.W.). Steam coal consumed on board ship.

BUNTON, or BUNTEN. See *Biat*.

BURDEN (Pa.). A charge of gunpowder, dynamite, &c., used in blasting coal or rock.

BURE (F., Belg.). A coal-pit.

BURGY (L.). Slack, or small coal.

BURNT-STUFF (M.). The contents of a *spoil bank* which has been thoroughly burned by spontaneous combustion. [A good material, when broken up and riddled, for stowing into the sites of *gob-fires*, and for packing in solid behind clay *dams* or *stoppings*.]

BURR (L.). Very compact siliceo-ferruginous sandstone.

BUSTER (really BURSTER). A machine for breaking down coals, &c., without the employment of blasting powder.

BUSTLE (Y.). Hurry in getting or working coal, or in performing other colliery work.

BUSTY (N.).

BUTTERFLY VALVE or CLACK. Pump valves constructed to open as shewn by dotted lines in Fig. 32. See *Bucket*.

BUTTOCK. That portion of a working *face* of coal, &c., next to be taken down.

BUTTOCKERS. Men who work at the *buttock*, or break out the coal ready for the *fillers*.

BUTTY. 1. (M.) A man who works a *stall*. He is a contractor, and performs or pays for the whole of the work done in *getting* and sending out the coal, &c., and keeping the *stall* in proper and safe working order. He sets the *timber*, *rips* the *gates*, *holes*, *packs*, fills coal into *tubs*, and is responsible to the *manager* for everything connected with his *place* (1), including the quality of the coal sent out. Sometimes as many as ten *butties* work a *stall*; they divide the money which is left over after paying the *holers*, *fillers*, and boys. They also pay for their own candles, smith's and carpenter's work, and find their own *picks* and other tools. Often termed a "Butty Collier." See *First Man*, *Joey*.

2. (M.) A man who sorts and fills into trucks, boats, &c., the coals upon the *bank* (1), for which he is paid by the ton. Known as a "Butty Banksman."

3. (M.) A mate, partner, friend, or fellow-workman.

BUTTYMAN (Y.). Contractors for getting coal, &c. See *Butty*.

BUTTYSHIP (S. S.). The prevailing mode of raising

the Ten-Yard coal seam. The contractor *gets, fills* in pit, and delivers coals to place of sale (masters finding timber, engine-power, and loaders into boats, &c.), finding all tools, horses, *skips*, corn, candles, powder, pit-beer, &c.

BUTTY SYSTEM (S. S., N. S., M.). When a pit is worked by contract, it is said to be worked upon the *butty system*.

BYARD. See *Biat*.

BYE CHAINS (S. W.). Hauling ropes (?) for *dip* inclined planes.

BYE-WORK (M.). Odd work, or that which is paid for by the day, in connection with the underground roads, &c. The men who perform it are called *Byeworkmen*.

# C.

CABIN. A small room fitted with wooden benches, a table, &c., in which the *Manager*, and other underground officials meet for consultation, writing reports on the state of the mine workings, having their *bait*, &c. In many large collieries there are several *cabins*, viz.—underviewer's cabin, men's cabin, lamp cabin, &c. Also on the pit bank there is always a *banksman's* cabin.

CAGE.—The apparatus in which the *tubs* of coal, the men, horses, and materials are raised and lowered in the *shaft*. *Cages* are constructed to carry from one to eight tubs or from 10 to 90 cwt. of coal, and are generally made of steel, and run up to 3½ tons in weight. A cage for holding four tubs is shown in Fig. 33.

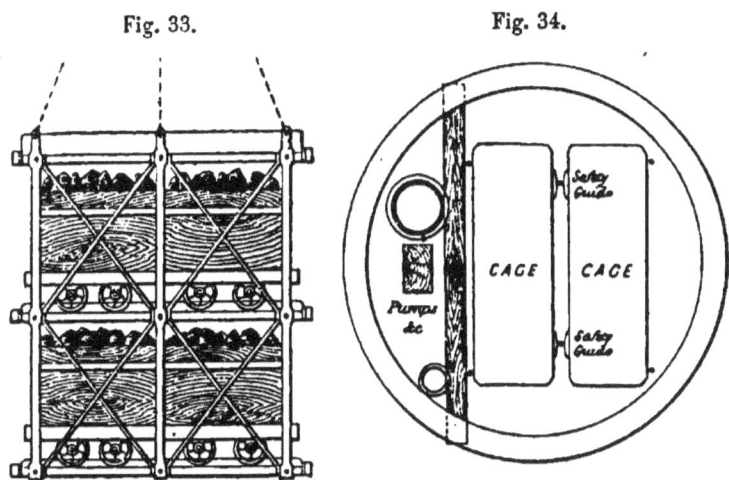

Fig. 33.   Fig. 34.

CAGE GUIDES. Vertical rods of pine, rails or rods of steel or iron fixed to *buntons* in *pit-shafts;* or wire cables fixed or suspended and weighted at pit bottom to prevent oscillation, between which the *cages* run, and whereby they are prevented from striking one another or against any portion of the shaft and the fittings contained therein. Fig. 34 is a plan showing a good arrangement of such guide when wire rope ones are used.

CAGE SEAT. Scaffolding, sometimes fitted with strong springs or with indiarubber blocks, to take off the shock, upon which the *cage* drops on reaching the *pit bottom*.

CAGING (N. S.). The operation of changing the *tubs* on a *cage*.

CAGE SHUTS (S.). Short props or catches upon which *cages* stand during *caging*. Fig. 35.

CAKING COAL. Coal of a bituminous nature, and has the property of agglomerating. It is not a free or open burning coal, and requires much poking on the fire.

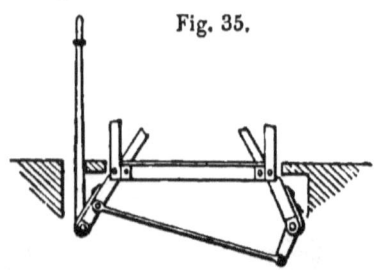
Fig. 35.

CALE (M.). A specified number of *tubs* taken into a working place during the *shift*.

CALING (M.). Conveying *tubs* into the *stalls* out of turn—irregularly—so that each is not supplied with an equal number during the day from each *train* or *set*.

CALLER (N.). A miner who goes round the villages two hours or so before work commences, to call up the men who first descend the pit to examine it in a morning.

CALLEY-STONE (Y.). A kind of *gannister*, which see.

CALLIARD or GALLIARD (N.). A hard, smooth, flinty grit-stone.

CALLOW. The *baring* or *cover* of *open workings*.

CALMSTONE (S.).

CANCH or CAUNCH (N.). That part of the roof of an underground roadway, which has to be taken down, or of the floor to be broken up, in order to equalize the gradient of such roadway. Fig. 36 is a diagram showing the bottom *canch b*, and the top one *a*, which are produced in consequence of the *fault slip* throwing the level of one roadway above the other.

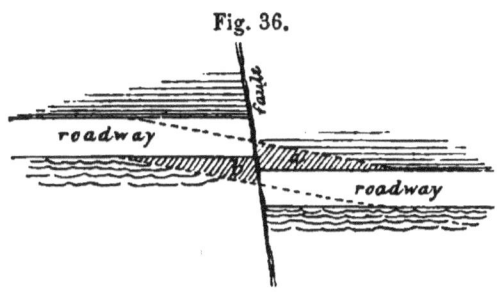

Fig. 36.

CANK or CANKSTONE (D. Lei.). See *Burr*.

CANKER. The ochreous sediment in coal-pit waters, being bicarbonate of iron precipitated by the action of the air upon that mineral.

CANNEL. A coal rich in hydrogen, produces much gas, and has a hard, dense structure. This word is derived from *Canwyl*, meaning a candle, from the readiness with which it lights and gives off a steady flame.

CANNON-SHOT. See *Blown-out Shot*.

CANNONIER (F.). See *Fireman*.

CANT. To slip or heel over to one side.

CANTEEN (N.). A small wooden barrel in which a collier takes his tea, &c., for refreshment during his *shift*.

CAP. 1. See *Blue Cap*.

2. See *Bar*.

Fig. 37.

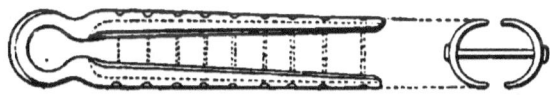

3. An attachment between a rope end and a chain, &c.: it is riveted on to the rope. See Fig. 37.

CAPPING. See *Cap* (3).

CAR. 1. (N. S.)  See *Canker*.

2. (Pa.) A *box* or tram (holds 75 to 140 cubic feet of coal).

CARBONATES. Black imperfectly crystallised form of diamond used for rock boring; the abrasion of the diamond removes the rock in an annular form, producing *cores*, which see.

CARRIAGE. See *Cage*.

CARROT. A solid cylindrical specimen or *core* cut in a *borehole* (1).

CART (Som., S. W.). A *tram* with or without wheels for conveying coals underground in *thin seams*.

CARTING (Som.). Hauling coals underground in *thin seams*.

CART TRADE (Som.). See *Land Sale*.

CARTRIDGES. 1. Paper or water-proof cylindrical cases filled with gunpowder, forming the charge for blasting. They are usually about 1¼ inches in diameter, and contain a quarter, half, and three-quarters of a pound of powder.

2. Short cylinders (about 4 inches long and 2½ inches in diameter) of highly compressed caustic lime made with a groove along the side, used in breaking down coal. See *Lime Cartridge*.

CARVING. 1. (Lei.) A wedge-shaped vertical cut or *cutting* at the *fast end* of a *stall*.

2. (Lei.) The *air-way* formed along the side of the *goaf* between the solid coal and a *pack wall*. See *Cutting*, Fig. 50.

CASE BOOK (N.). A book kept at a colliery in which the name and description of every horse or pony which is off work for 24 hours or longer, and the driver's name, is entered. It is examined periodically by the *viewer*, the reason and cause of every animal being off work being fully enquired into.

CASH (S.). Soft shale or *bind*.

CAT, or CATCH-EARTH (S. S.). A *clunchy* rock.

CATCHER. 1. A safety or disengaging hook for *overwinding*.

2. (L.) See *Cage Shuts*.

3. Very strong beams in *pit-shafts* (of oak or wrought iron) to catch the rods, &c., of pumps in case of a break down, to prevent them falling downwards.

CATCHES. 1. Iron levers or props at the top and bottom of a *pit shaft*. See *Cage Shuts*.

2. Iron stops fitted on a *cage* to keep *trams* from running off.

3. Projecting blocks of wood attached to pump *spears* for preventing damage in case of a break down.

CATCH SCAFFOLD. A platform or cradle in a *pit-shaft*, placed a few feet beneath a working scaffold in case of accident.

CATHEADS (N.). Nodular or ball ironstone.

CATRAKES. Cataracts of a Cornish pumping engine, first introduced by Boulton and Watt.

CAVILLING RULES (N.). Rules or bye-laws in reference to *cavils* and wages.

CAVILS (N.). Lots, drawn for quarterly by *hewers* for every working place in the pit: in the *broken* or in splitting *pillars*, one *pillar* equals a *cavil*.

CAULDRONS (S. W.). See *Bell Moulds*.

CAUM (CUM.).

CERTAIN RENT. See *Dead Rent*.

CHAIN-BROW WAY. An underground inclined plane worked by an *endless chain*.

CHAIN ROAD. An underground *wagon-way* worked upon the *endless chain* system of haulage.

CHAIR. See *Cage*.

CHALK and PIPE-CLAY (N.). An expression used by *sinkers* and *borers* for gypsum.

CHAMBER AND PILLAR (Pa.). See *Breast and Pillar*.

CHALDER WAGON (N.). A railway truck holding 53 cwt. of coals.

CHALDRON (N.). An ancient measure (*Chalder*) equal to 2000 lbs., but 53 cwt. is now customary, though seldom used.

CHALKING-ON (N.). Keeping an account of the number of *tubs* sent out of a *stall*, &c.

CERTIFICATED MANAGER. See *Manager*.

CHANCE MEASURE. Any seam or bed of coal or other rock occupying an unusual or foreign position in the strata.

CHANGER AND GRATHER (N.). A man whose duty it is to keep the pump *buckets* and *clacks* in working order about a colliery.

CHAP. 1. (S.) A customary and rough mode of judging from the sound, of the thickness of solid coal existing between two places near to each other. The sound is produced by knocking with a hammer on the solid coal.

2. (S.) To examine the face of the coal, &c., for the sake of safety, by knocking on it lightly.

CHARGEMAN (M.). A man specially appointed by the *manager* to fire *shots* and to look after the *blowers* (2).

CHARGEUR (Belg.). A woman or girl who loads coal into *trams* in the mine.

CHARTER (M.). A price per ton paid to *butties*.

CHARTER MASTER. Head *butty* or contractor.

CHECK. A *fault*, which see.

CHECK-WEIGHMAN. A man appointed and paid by the *colliers* (1) to weigh the coals on reaching the surface. He must have been employed in the mine, and must not interfere with the ordinary weighman.

CHEEK. A projecting mass of coal, &c.

CHEESES (D.). Clay ironstone in cheese-shaped nodules.

CHEMIST'S COAL (S.). An ancient term given to a particular kind of hard *splint* coal which used to be carried by women in their *shifts* or *chemises* out of the mines. The word *chemise* became changed into *chemists*.

CHERKERS (F. D.). See *Catheads*.

CHERRY COAL. A soft, velvet-black, caking, bright resinous coal.

CHEST (S.). A tank or barrel in which water is drawn from the *sump*.

CHIMNEY. A spout or pit in the *goaf* of vertical coal-seams.

CHIMNEY WORK (M.). A system of working a great thickness of beds, or *pins* of *clay ironstone*, in patches or areas of from 10 to 30 yards square, and 18 or 20 feet in thickness. The bottom beds are first worked out, and then the higher ones, by the miners standing upon the fallen débris; and so on upwards in *lifts* (3). See *Rake*. See Fig. 38.

Fig. 38.

CHINGLE (S.). Portion of the coal-seam used for *stowing* purposes.

CHINKS (S.). Holes in *brattices*.

CHITTER. 1. (L.) A seam of coal overlying another one at a short distance.

2. (D.) A thin band or *pin* of *clay ironstone*.

CHOCK. A square pillar constructed of short rectangular blocks of hard wood, for supporting the *roof*.

Fig. 39.

Fig. 40.

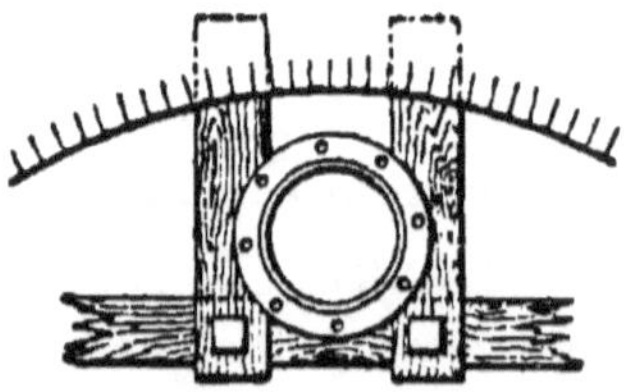

They are generally built upon a few inches of slack, or rubbish. See Fig. 39.

CHOGS (Y.). Blocks of wood for keeping *pump-trees* or other vertical pipes plumb. See Fig. 40.

CHOKE DAMP. See *Black Damp*.

CHOP (Som.). See *Fault*.

CHUMP. To drill a shot-hole by hand.

CHURNS (F. D.). Ironstone workings in cavern-shaped excavations. A kind of rough *chamber and pillar* system of working.

CHUTE (Pa.). A *bolt* or *thirl* connecting a *gangway* with a *heading*.

CINDER COAL. Coal near to a *trap* or *whin dyke*, of altered nature, due to the heat of the lava.

CIRCLES (Ch.). Wavy, undulating lines of various colours frequently seen in the sides of *shafts*, on the *pillars*, *faces*, and *roof* of rock-salt mines. They vary from a few feet to a few yards across, and are caused by the form of the stratification of the rock salt, which is usually spheroidal, or wavy and undulating, being cut through or dressed to a plane.

CIRCLE SPOUTS. See *Garland* (1).

CLACK. The lower valve of a lifting or forcing *set* (1) of pumps, made something like a *bucket*, without the central rod.

CLACK-DOOR PIECE. A cast-iron pipe, having a doorway made in the side of it for giving access to the *clack*. The *clack-door* is an iron plate bolted to the *door-piece*.

CLAGGY. Sticky.

CLAMS or CLAMMS. Strong iron clamps for firmly holding pipes, ropes, &c., in *shafts*, or on inclined planes.

CLANNY. A *safety-lamp*, the invention of one Dr. Clanny. First exhibited in Sunderland in the year 1813. The lower part of the lamp-top around the flame is constructed of a thick glass ring, above which is the wire gauze chimney. It is a lamp which gives a good light, and indicates freely the presence of firedamp, but is not so safe a lamp as some others.

CLAY. In mining language usually means tender shale, or indurated clay.

CLAY BAND (S. W.). Argillaceous ironstone in thin beds, very numerous in the lower *coal measures*.

CLAY DAM. 1. (M.) A *stopping* made of puddled and well-beaten clay, from 12 in. to 36 in. thick, and well rammed into the roof, floor, and sides of the excavation made to receive it.

2. A *stopping* consisting of two walls of stout planks placed 18 to 24 inches apart, and supported on the outsides by upright props; good strong clay well beaten and puddled into the space between the walls of planks forms a tolerably strong barrier against water pressure.

CLAY-HOG (M.). Kind of *wash faults*, or *lows*. See Fig. 70 (No. 2).

CLAYING. Lining a *borehole* (2) with clay, to keep the powder dry.

CLAYING IRON. See *Bull* (1).

CLAY-IRONSTONE. A dull brown or black compact form of siderite, with a variable mixture of clay, and usually also organic matter. Occurs in the carboniferous and other formations in the form of either nodules, where it has usually been deposited round some organic centre, or of beds interstratified with shales and coals.

CLEADING. Deal boarding for *bratticing* or *lagging*.

CLEAN. 1. (N.) Free from *firedamp* or other noxious gases.

2. A coal-seam is said to be *clean* when it is free from *dirt partings*.

CLEANSER, or CLANSER. An iron tube or shell, with which the *bore-meal* is extracted from a *bore-hole* (1).

CLEAR. See *Clean*.

CLEARERS (I.). Colliers who *hole* the coal, working at distances of say three or four yards apart along the *face*.

CLEAT. 1. Natural jointing of coal seams, with generally a north and south direction, irrespective of *dip* or *strike*.

2. (M.) A wooden wedge four or five inches square placed between the head of a *puncheon* and the underside of a *bar* or *cap*.

CLEATS (N.). A system of natural joints or fissures running through the great northern coal-field of Durham, &c., ranging N.N.W.

CLEAVINGS. Horizontal divisions of beds of coal, &c., or in the direction of the laminæ.

CLEEK. 1. (S.) To load *cages* at the *pit-bottom*, or at *mid-workings*.

2. (S.) A *haulage clip*.

CLIFF or CLIFT (S. W.). Shale which is laminated, splitting easily along the planes of deposition. See *Bind*.

CLINKER. See *Cinder Coal*.

CLIP. See *Haulage Clip*.

CLIP PULLEY. A wheel containing clips in the groove for gripping a wire rope.

CLIVVEY. A ⌒-shaped iron ring, by which a chain is attached to a rope *cap* (3).

CLOD (D. Lei.). Indurated clay, not flaky.

CLOD-TOPS (F. D.). Overclays, or clayey beds overlying seams of coal.

CLOG-PACK (Y.). See *Chock*.

CLOGS (M.). Short pieces of timber about 24" × 6" × 3" fixed between the *roof* and a *prop*.

CLOSE WORK. 1. *Driving a tunnel*, or *drifting* between two coal-seams.

2. (S.) See *Narrow Work*.

CLOSING APPARATUS. Sliding-doors or other mechanical arrangement at the top of an *upcast shaft* for allowing the *cages*, &c., to pass up and down without disturbing the *ventilation* of the mine. Fig. 41 shows a side elevation of a self-acting arrangement, in which horizontal iron doors or slides are actuated by long levers or arms worked to and fro by the cages.

Fig. 41.

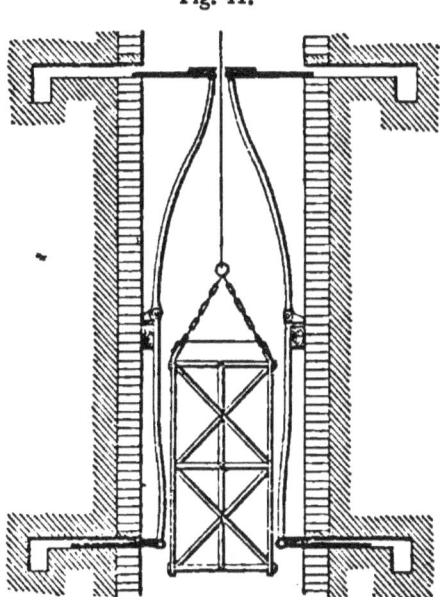

CLOT. See *Clod*.

CLOTHING. *Brattice* constructed of a coarse canvas specially prepared.

CLUMPER (F. D.). A large mass of fallen stone in the mine.

CLUNCH (M.). A kind of hard earthy *Fireclay*.

COAL. 1. All vegetable matter which has been changed under the influence of ages of time, and which is capable of undergoing combustion in contact with oxygen. It is fossil fuel—fuel produced and stored up in bygone ages, which by chemical and physical agencies, with and without the presence of heat and moisture, has been modified or resolved into the various forms which bear this name. It is a compact black rock or mass, having a fracture usually of resinous lustre, usually friable, inflammable, burning with flame, smoke, and smell. The substance of coal is principally carbon, viz., 74 to 97 per cent. The sp. gr. varies between 1·3 and 1·5. The weight of a cubic foot of solid coal equals 74 to 82 lbs.; heaped coal from 45 to 55 lbs. It occupies from 40 to 50 cubic feet per ton in the heaped or broken state. It occurs in beds or seams intercalated between strata of shale clay, sandstone, &c., in geological formations of Palæozoic, Secondary, and Tertiary age. The thickness of coalseams ranges from mere *sheds* (3) to between 100 and 200 feet.

2. Coal in large lumps, as distinguished from *slack* or *small*.

COAL BEARING (S.). The ancient custom of employing women to carry out on their backs the produce of the mine.

COAL BED. A formation in which there are one or more strata of coal: the stratum or strata of coal themselves.

COAL BRASSES (S. W.). Iron pyrites in coal seams.

COAL BREAKER (Pa.). Machinery consisting of iron rolls, shoots, and *screening* apparatus for preparing *anthracite* for the market.

COAL-CUTTING MACHINE. An engine with mechanism combined, generally worked by compressed air, for *holing* or *undercutting* a seam of coal.

COAL-DRAWING. The operation of raising or *winding* (1) coals at a colliery.

USED IN COAL MINING, ETC. 55

COAL-DROP. Broad shallow inclined trough, down which coals are discharged from waggons into the holds of *colliers* (2) and other vessels.

COAL DUNS (F. D.). *Coal measure* shales, &c.

COAL DUST. Very finely-powdered dust suspended in the air-currents in mines, composed of coal and other finely-divided substances. It is capable of extending and aggravating an *explosion* of *fire-damp*. When mixed with even less than 1 per cent. of this gas, an explosive mixture is obtained under certain conditions.

COAL FACE. The *working face* or *wall* of a *stall*, composed wholly of coal.

COAL-FIELD. A district containing *workable* mines of coal; generally applied to areas composed chiefly of the *coal measures*, though rocks of more recent date may overlie them, or they may be partially submarine. The thickness of some coal-fields is very great, that of Saarbrucken in Germany being 20,000 feet, South Wales, 14,000 feet. The number of separate coal-fields in England is sixteen, Scotland six, Ireland five, covering an aggregate area of something like 5000 square miles. The following figures represent the total thickness of *coal measures* and of the various *coal beds* contained therein, in some of the principal districts:—

| Coalfield. | Feet. | | Feet. | |
|---|---|---|---|---|
| North of England | 2,100 | of measures, | 50 | of Coal. |
| Midland | 3,000 | „ | 45 | „ |
| Scotland | 4,344 | „ | 95 | „ |
| Lancashire and Cheshire | 7,000 | „ | 70 | „ |
| N. Staffordshire | 5,000 | „ | 140 | „ |
| S. „ | 1,800 | „ | 50 | „ |
| Warwickshire | 3,000 | „ | 26 | „ |

| Coalfield. | Feet. | | Feet. | |
|---|---|---|---|---|
| Leicestershire | 1,800 | of measures, | 45 | of Coal. |
| Bristol and Somerset | 5,000 | ,, | 81 | ,, |
| Forest of Dean | 2,300 | ,, | 27 | ,, |
| South Wales | 10,000 | ,, | 179 | ,, |
| Ireland | 1,800 | ,, | 17 | ,, |
| Prussia | 7,218 | ,, | 294 | ,, |
| Pennsylvania | 2,750 | ,, | 70 | ,, |
| India | 12,000 | ,, | 350 | ,, |
| China (10,000 sq. miles) | — | ,, | 40 | ,, |

Although Great Britain has during the last thirteen or fourteen years been producing over 100,000,000 tons of coal annually (156,500,000 during 1882) from about 3,800 collieries, it has been estimated that there remains something like 135,000,000,000 tons still available, which includes all coal seams above 2 feet in thickness to a depth of 4000 feet, after deducting 40 per cent. for loss and other contingencies.

COAL-GETTER. One who *cuts, holes, hews,* or *blows* coal in the mine.

COAL HAGGER (N.). One who is employed in cutting or hewing coal in the pit.

COAL HEUGHS (S.). Mounds of refuse about old pits. They date as far back as 1545.

COALING (M.). Engaged in *cutting* [see *Cut* (2)] and *getting* coal.

COAL-MASTER. The owner or lessee of a coal-field or colliery, who works it and disposes of its produce.

COAL MEASURES. The upper division or *series* of the *carboniferous system* of rocks, containing almost exclusively the whole of the coal of the earth.

COAL PIPE. 1. The carbonised annular coating or bark of a fossil plant.

2. A very thin seam of *shed* of coal.

COAL PRINTS (N.). Thin films or patches of coal-like matter interbedded with shale, &c.

COAL-RAKE (D.). A seam or bed of coal.

COAL ROAD. An underground roadway or *heading*, made or *driven* entirely within the seam, or one having a coal *roof* and *floor* as well as coal sides.

COAL SALAD (S. W.). A mixture of various sorts of coal.

COAL SEAM. See *Coal Bed*.

COAL SHALE (F. D.). See *Coal Measures*.

COAL SHED. A bed of coaly matter only a few inches in thickness, and therefore unworkable.

COAL SMITS (Y.). Worthless, earthy coal. See *Coal Smut*.

COAL SMUT. A black, earthy coaly stratum at or near the *surface*. The *outcrop* of a coal seam.

COAL-STONE. A kind of *cannel*.

COAL WARRANT (N. W.). A kind of *clunch* or *fire-clay* forming the floor of a coal seam.

COAL WASHING. See *Washing Apparatus*.

COAL WORK (N.). *Headings*, &c., driven in a seam of coal.

COB (D.). A small solid pillar of coal left in a *waste* as a support for the *roof*.

COBBLES. *Round coal* in smallish lumps.

COBBLING. Cleaning the roads in the pit of coals which have fallen off the *trams* during the *turn* (1).

COCKERMEGS. Timber props fixed in the manner shown in Fig. 42, to support the coal during *holing*.

Fig. 42.

COCKERPOLE. A piece of timber placed horizontally between two inclined pieces which abut against the *roof* and *floor*. See Fig. 42, *Cockermegs*.

COCKERS. See *Cockermegs*.

COCKERSPRAGGS. See *Cockermegs*.

COCKHEAD (D.). A description of *pack* or support to the roof of a *waste*, consisting of a *gobbin* of slack or rubbish about 12 feet in width, surmounted by a few lumps of coal.

COFFERING. Watertight casing or walling of a *shaft* without the employment of metal *tubbing*. It consists in lining the *shaft* to stop the influx of *feeders* of water where the *head* of water is not great by means of brickwork set in hydraulic mortar backed with puddled clay or with soil; the water being allowed to escape down a wooden pipe called a *plug-box* during the putting in of the *coffering*.

COG. 1. See *Chock*.
2. (S. S.) A *pack*, which see.

COG AND RUNG-GIN. One of the earliest appliances

for raising the coals and water from coal pits. It was a kind of windlass fitted with a cog-wheel and pinion arrangement, and worked by a horse in much the same way as our nineteenth century *horse-gins* are worked.

COGGER. One who builds up *cogs* (1) (2).

COGGING (S. S.). The propping up of the roof in *longwall stalls*.

COKE-COAL (N.). Carbonised or partially burnt coal found on the sides of *whin dykes*.

COKING COAL. A coal having the property of being converted into large and hard cokes, free from sulphur, &c.

COLD FURNACE (N.). A *drift* driven up into an *upcast shaft* to convey the *return air* into it instead of passing it over the *furnace* fire. This is done to guard against any *gas* in the *return air firing* (3) from the heat of the *furnace*.

COLD PIT (Lei.). A *downcast* pit. Called *cold* because the fresh or cold air comes down it.

COLLAR (N.). The mouth of a pit-shaft.

COLLAR-CRIB (N.). A strong oak polygonal frame fixed in a shaft, upon which the wooden *wedging crib* of solid wood *tubbing* is bedded.

COLLARING. Timber framing for steadying and supporting *pump trees* in a shaft. See Chogs, Fig. 40.

COLLIER. 1. Strictly speaking, a man who cuts or hews coal with a *pick*, though commonly applied to any one who works in or about a colliery.

2. A steam or sailing vessel carrying a cargo of coals from *staithes* and *drops* (2) coastwise.

COLLIER'S COALS. A certain weight of coals allowed periodically (once in a month or six weeks) by the owners to the *colliers* (1) and other men employed on the works, who are in most cases householders, as a perquisite. The colliers, however, are not as a rule paid for cutting and hauling these coals.

COLLIER'S (1) TON. A weight of often several cwt. in addition to the standard ton or 2240 lbs. In former times as much as 28 cwt. was reckoned as one ton.

COLLIERY. A place where coal is mined, with its machinery and *plant*.

COLLIERY CONSUMPTION. The amount of fuel consumed in generating steam and for other purposes in and about a colliery establishment.

COLLIERY WARNINGS. Telegraphic messages despatched from the Government meteorological stations to the principal colliery centres to warn the managers of mines when any sudden fall of the barometer is taking place, in order that extra vigilance and care may be taken in guarding against the effects of possible sudden outbursts of *fire-damp,* or of unusually large quantities of that gas being given off from old workings, &c., as a consequence of a reduced atmospheric pressure.

COLUMN. 1. The rising main (either fixed vertically or inclined) or length of pump-trees or pipes conveying the water from the mine to the surface.

2. *Ventilating column,* which see.

USED IN COAL MINING, ETC. 61

3. See *Carrot*.

COMB COAL.

COME (Come Water). The constant or regular flow of water in a mine proceeding from old workings or from watery rocks.

COMET (S. W.). An open-burning hand lamp with a long torch-like flame.

COMING UP TO GRASS or COMING UP TO DAY. A common term used by miners for the word *Basset*.

COMPANY. A number of *butty* colliers who work and carry on a *stall*, &c.

COMPOUND VENTILATION (N.). The system, first practised by Buddle, of dividing up or *splitting* the *air*, and of ventilating the workings of a coal mine by giving to each *district* or *panel* a separate quantum of fresh air, and conveying away the *return air* to a main *return* direct from each *panel*.

CONDUCTORS. See *Cage Guides*.

CONE-IN-CONE COAL. Steam or anthracite coal exhibiting a peculiar fibrous structure passing into a singular toothed arrangement of the particles called cone-in-cone coal or crystallised coal.

CONICAL DRUM. The *rope roll* or *drum* of a *winding engine* constructed in the form of two truncated cones placed back to back, the outer ends or sides being usually the smallest in diameter. See Fig. 43. The winding ropes are wound and unwound in a spiral form, and rest in channels or grooves of iron riveted upon the lagging. Drums of this description are in use chiefly at deep pits where a large *output* is required and a high speed of *winding* is a necessity. They range from say 12 feet to 32 feet in diameter, and, together with the main shaft, weigh as much as 60 tons. The object of the spiral or scroll form is to equalise the load upon the engines at all points during the *lift* or run, without the employment of any special balancing arrangements, such as chains, &c.

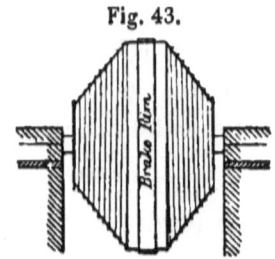

Fig. 43.

CONSEY (S.). A branch underground road in *stoop and room* workings.

CONVERTING COAL (M.). A local name given to a coal suitable for steel-making purposes at Sheffield, &c.

COOMING (S.).

CORES. The cylindrical-shaped samples of strata produced by the *Diamond system* of *boring* (1). They vary in diameter from 1 to 18 inches, and are obtained whole in lengths of many feet under favourable circumstances.

CORF-BATTER or CORF-BITTER (N.). A lad who cleans the dirt or mud off *corves*.

CORF, CORFLE, or CORVE (N.) (from the Dutch *Korf*, a basket). See *Box*. But when used for bringing up the débris from a *sinking pit* they are made without wheels, and are more like a basket. In bygone days *corves* were wicker baskets, having wooden *bows* or handles: they held about 4½ cwt. of coal.

CORNERS (S. W.). Bands of *clay ironstone*.

CORNISH PUMPS. Pumps arranged and worked upon a system very common in Cornwall, and very frequently applied to colliery drainage. The system consists in having a *lifting* pump at the bottom of the pit to raise the water out of the *sump*, and a series of *force* pumps, placed one above another, to drive it up by stages to the surface or *adit*, the whole of the pumps being worked simultaneously from the main rod.

CORPORAL (M.). An overlooker of the pony boys and others upon the underground *ways* in a *district*.

CORROIS (F.). Clay or *wax dams* and *walls* built up to isolate the place of a *gob-fire*.

CORVERS (N.). Carpenters who make *corves*. A *corver* was formerly paid 4½d. per score of *corves* brought up out of the pit, being bound to find the pit in *corves* and keep all in repair.

COUNTER CHUTE (Pa.). An empty, or worked out *breast*, down which coals are *dumped* to a lower level, or *gangway*.

COUNTER COAL (Pa.). Coal worked from *breasts* or *boards* to the *rise* of a *counter gangway*.

COUNTER GANGWAY (Pa.). A *level* or gangway driven at a higher level than the bottom of the shafts, or foot of the *slope*.

COUNTER HEAD (M.). An underground heading driven parallel to another, and used as the *return air course*.

COUNTRY PITS (F. D.).

COUP (N.). To exchange *cavils* with the consent of the *overman*.

COUPLE (M.). To conduct water which runs down the sides of shafts into water *curbs* or *garlands* (1).

COUPLING (Y.). The *cap* (3) of a rope.

COUPLINGS. See *Double Timber*.

COURSE. 1. To conduct the ventilation of the colliery backwards and forwards through the workings, by means of properly arranged *stoppings* and *regulators*.

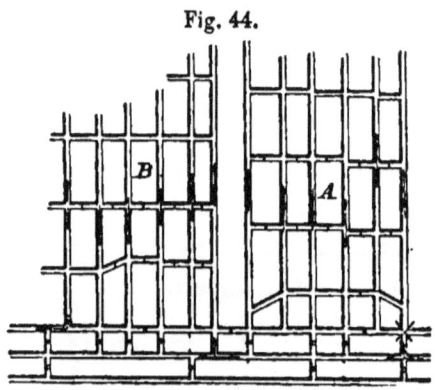

Fig. 44.

In Fig. 44, which gives a plan of two panels, or blocks of *board and pillar* workings, that set marked A shows the system of *coursing* known as *two and two*, whilst in B, the workings are *coursed three and three*; that is to say, the ventilation is conducted up and down two and three *boards* respectively, as indicated by the darts.

2. (Som.) A seam of coal.

COURSING THE WASTE. Threading the *ventilation* up certain *workings* and down others.

USED IN COAL MINING, ETC. 65

COVER (N.). The total thickness of strata overlying the workings of a seam of coal, &c. If a mine is 1800 ft. deep at the shafts, the *cover* will be 1800 ft., but if the workings are level and extend underneath rising or falling ground at the surface, then the *cover* will be greater or less as the case may be.

COVERING BOARDS (Y.). A series of *boards* and *thirls* formed on the side of a *shaft pillar*, out of which *long-wall* working is commenced on No. 1 method. See Fig. 92, *Long-wall*.

COW (N.). See *Backstay*.

COWLS (N.). Wrought-iron water-barrels, or tanks, attached to the winding ropes, and emptied at the surface, used when the engines are not *winding* (1) coals.

CRACKS (S.). Vertical planes of cleavage in coal, &c., running at right angles to *backs*.

CRACKET (N.). A tool used by colliers in getting coal.

CRADLE. 1. A moveable platform or scaffold suspended by a rope from the surface, upon which repairs or other work is performed in a shaft.

2. (M.) A loop made of a chain in which a man is lowered and raised in a shaft not fitted with a cage.

CRANE BOARD (N.). A *return air course* connected directly with the *furnace*.

CRANK (N. W.). Small coal.

CREASE (F. D.). ⋅ Mountain limestone of ironstone workings.

CREEL (S.). A kind of basket in which coals and

F

débris were conveyed from the pit. They were carried on the backs of *bearers*, being steadied by a strap round the forehead.

CREEP. 1. The gradual upheaval of the *floor* of a mine towards the *roof*, due to the weight of the *cover* and a tender floor. The working away of a seam of coal will often produce *creep* in an underlying seam, as well as a corresponding subsidence or creep in one overlying it at no great distance. See Fig. 45.

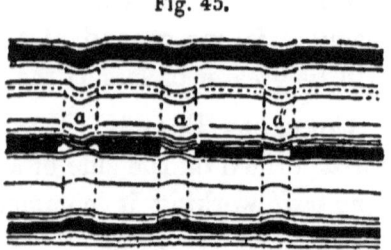

Fig. 45.

2. A very slow movement of a winding engine, when the brake is not sufficiently applied to hold it quite fast.

CREEPING. The settling down, or natural subsidence, of the surface and buildings, &c., thereon, caused by the extraction of mines to such an extent as to produce such settlement. Workings shown in Fig. 16 will not create any *creeping* of the surface, but as soon as the *posts* or *pillars* are worked away a subsidence may be expected, the extent of which will depend upon the depth to the coal worked, its thickness, dip, the nature of the overlying *measures*, and the way in which the *building* or *stowing* is done.

CREESHY (GREASY) BLEAS (S.). Nodules of bituminous shale met with in the soft roofs of some of the Scotch collieries. So called from the sort of unctuous smoothness, which causes them to fall out when the coal is worked away from beneath them.

CREPT-BOARDS. *Boards* more or less filled up from the effects of *creep*. See Fig. 45, a a' a".

CRESSET. A *fire-lamp*, which see.

CRIB. 1. A cast-iron ring in a shaft upon which *tubbing* is built up. See *Wedging Crib*.

2. A wood ring upon which the brick lining or *walling* of a shaft is built. It is constructed in segments (six or eight to the circle) which are bolted together as shown in Fig. 46, which gives a plan and elevation of one segment with joint blocks and bolts complete.

Fig. 46.

CRIBBING (N. E.). See *Tubbing*.

CROOK (B.). A self-acting apparatus for running the *hudges* on inclines in *steep seam* workings.

CROP. 1. See *Outcrop, Bassett*.

2. The roof coal or stone which has to be taken down in order to secure a safe roof in the workings.

CROPPER. A *shot* placed at the edge or rise side in a *sinking pit* bottom.

CROSS (S. W.). See *Cross-cut* (2).

CROSS-CUT. 1. A drift or heading driven through or across the *measures* from one coal seam to another. See *Branch*.

2. A headway which is driven at an angle to the vertical planes of cleavage.

CROSS GATES (Y.). Short headings driven on the *strike* right and left out of and at right angles to the main *gates*.

F 2

CROSS-HOLE (S. W.). A short *bolt hole* or *cut through* communicating with two *headings*, for ventilation purposes.

CROSSING. 1. See *Air crossing*.
2. (N. W.) A *Cross-cut*.

CROSS-MEASURES. A line drawn horizontally or nearly so, through or across inclined strata: e.g. a *branch* or *crutt* is a *cross-measures drift* or *heading*.

CROSS OFF (Cl.). See *Stack out*.

CROW COAL. See *Anthracite*.

CROWN IN (Ch.). The surface or *cover* of a rock salt mine is said to *crown in* when it falls in or produces *creep*.

CROWN or CROWN-TREE (N.). See *Bar*.

CROWNINGS IN (S. S.). The strata forming the *roof* or *cover*.

CROW'S FOOT. An iron claw or fork, forming part of the boring tackle for deep *boreholes*, to which a rope is attached, and by which the rods are lowered and raised when changing the cutting tools, &c. See Fig. 47, which is called an open *runner* (3).

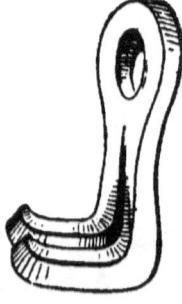

CROW-STONE (D. Y.). See *Gannister*.

CROZLE (D.). To cake or harden.

CROZZLING. Aggregation of coal when burning.

CRUSH. The breaking up or *weighting* of *pillars* of coal due to the pressure of the overlying rocks and to the hardness of the *floor*.

CRUST (Sh.).* Whitish fine sandstone.

CRUTT (N. S.). See *Branch*.

CRYS GROUND (F. D.). Carboniferous limestone strata containing beds of iron-ore.

CUBE (S.). See *Furnace*.

CUBE COAL. Coal broken up into cubes of about one foot square to suit the trade.

CUFFAT (F.). A vessel in which coals are sometimes raised in the shaft, consisting of a kind of shallow tub fitted with 4 wheels and attached to chains at the sides, the coals being piled up in a conical form and kept from falling off by iron rings placed round them one above another. See Fig. 48. Some Cuffats are made as much as 9 feet deep and more like the English *Bowk*.

Fig. 48.

CUILLER (F.). A long wrought-iron cylindrical bucket in which the débris made by the boring in the *kind-chandron* system of shaft sinking, is brought to the surface. Whilst the larger of the two cutting tools employed in boring out the shaft is at work, the *cuiller* remains in the bottom of the small bore in the centre of the shaft, which it nearly fits, and catches the stuff as it falls from the upper or fully bored out portion of the pit. Are made up to 12 tons capacity.

CULBUTEURS (Belg.). *Tippers* which turn completely over or round.

**Culm** (S. W.). Inferior *anthracite*, and the small or slack of smokeless coal. The Kilkenny coal of Ireland.

**Cundie** (S.). The spaces from which coal has been worked out, partially filled with dirt and rubbish between the *buildings* or *packs*. See *Waste*.

**Cupola.** 1. The offtake for smoke and *return* air erected at or near to the top of the *upcast shaft*.
2. See *Furnace*.

**Curb.** See *Crib*.

**Curb Tubbing.** Solid wood *tubbing*.

**Curbing.** See *Back-casing*.

**Curf** (Som.). The *floor* of an underground way which is being taken or broken up. See *Caunch*.

**Curley Cannel.** Cannel coal which breaks with a conchoidal or curly fracture. It is often used for oil manufacture.

**Curl-stone** (Sh.). Ironstone exhibiting *cone-in-cone* formation.

**Curry-pit** (Lei.). A hole or very shallow pit sunk from an upper to a lower portion of a thick seam of coal through which the *return air* passes from the *stalls* to the *air way*, which is carried alongside and parallel to the side of the stalls, and sometimes underneath the *goaf*. See plan, Fig. 49.

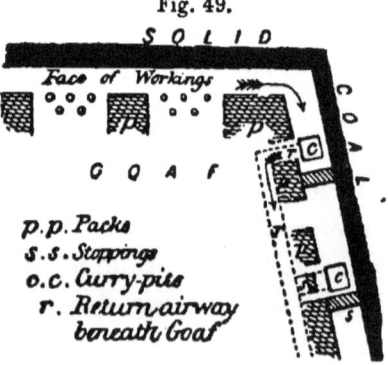

Fig. 49.

p.p. *Packs*
s.s. *Stoppings*
o.c. *Curry-pits*
r. *Return airway beneath Goaf*

CUT. 1. (Som.) A *staple* or *drop-pit*, which see.
2. To hew or hack coal, &c., with a *pick*.
3. (S.) See *Buttock*.
4. The depth to which a drill hole is put in for blasting.

CUT-CHAIN (S.). A system of working underground self-acting inclined planes from several different levels communicating with such incline, by means of chains of various lengths which are regulated according to the level from which it is intended to lower the coals.

CUT-OUT. 1. (F. D.) See *Crutt* or *Branch*.
2. When a fault which dislocates a seam of coal more than its entire thickness, the seam is then said to be *cut-out*.

CUT-OVER (M.). To cut or nick the seam of coal in a *long-wall* working, over or beyond the first joint or *cleat*, running more or less parallel with the *face* line. This is done in order to extract the coal in as large lumps as possible without the use of powder and with a minimum of labour in *getting*.

CUT-THROUGH (N. S.). *Bolt-holes* put through between headings every 18 to 20 yards in mines having a steep inclination. See *Dip* (4). Fig. 54.

CUTTER. 1. (S.) A fissure or natural crack in strata.
2. (Pa.) Joints at right angles to *backs*.

CUTTING-OFF ROAD. A slant road in *long-wall* workings, out of which the *stall-gates* are branched parallel to the *main road*, and which at certain distances cut off a range of *stalls* to the rear. See *Long-wall*, Fig. 92.

CUTS (S.). Strips of coal worked off the sides of *pillars*.

CUTTING. The end or side of a *stall* next to the solid coal, where the coal is cut with a pick in a vertical line to facilitate breaking down. See plan of a cutting, Fig. 50.

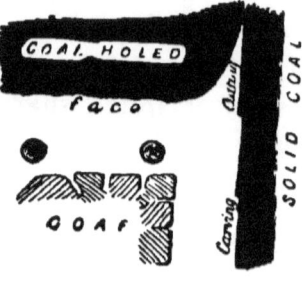

Fig. 50.

CUT-UP (S.). The breaking down of the roof to a considerable height.

CUVELAGE (F.). *Tubbing*, which see.

## D.

D. C. *Down-cast* (1), which see.

D LINK. A flat iron bar attached to chains, and suspended from a hemp rope to a windlass at surface. It is a loop in which one man is lowered and raised in an *engine-pit*. He sits upon the flat bar, the chains passing up in front of him, and the leather strap or belt is fastened round the back under the arms. See Fig. 51. He is free to move his legs and arms, and to turn himself about in any direction, and to perform work with a spanner or hammer, &c. Fig. 52 is a sketch of a form of hook commonly used for suspending the D link to the rope.

Fig. 51.   Fig. 52.

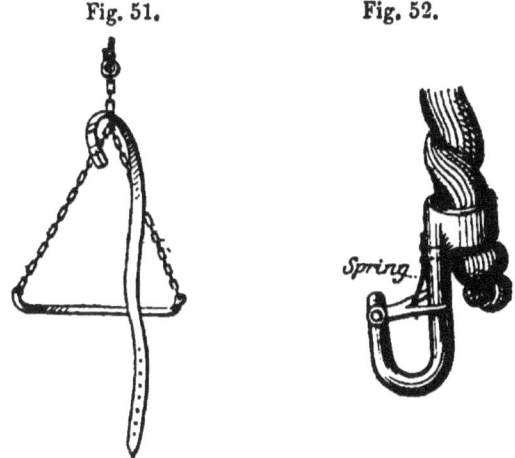

DADDING (N.). Mixing *firedamp* with atmospheric air to render it incapable of ignition. See *Brush* (1).

DAM. 1. An underground *stopping* or wall constructed of masonry or of clay, by means of which gas or *damp*, and spontaneous combustion, are prevented from escaping and breaking out.

2. A solid brick or timber stopping for keeping back accumulations of water.

DAMP. (From the German, Dampf.) Carbonic acid gas, or a mixture of gas (fire-damp) and air, incapable of supporting combustion, and therefore unfit for respiration.

DAMPED. Suffocated by gas or foul air in a mine.

DAMPY (M.). A pit is said to be *dampy* when the *air* in it is mixed with so much carbonic acid gas as to cause the lights to burn badly or to go out.

DAN. 1. (M.) A tub or barrel, sometimes with and sometimes without wheels, in which mine water is conveyed along underground roadways to be discharged into the *sump* or *lodge*, or raised in the *cage* to the surface.

2. A small *box* or sledge for carrying coal or *débris* in a mine.

DANGER-BOARD. See *Fire-board.*

DANKS (S.). See *Bat.*

DANT (N.). Sooty, worthless coal.

DANTY (N.). Disintegrated coal.

DARG (N.). A specified quantity or weight of mineral agreed by masters and men to be worked during a *shift* for a certain sum of money.

DASH (N.). See *Dadding.*

DATALLING. Blowing down *roof* in a mine.

DATLERS (L.). Men who work underground, not being contractors, and are paid by the day.

DAUGH (S.). *Underclay,* or *holing dirt.*

DAVY. A *safety lamp*, invented by the late Sir Humphrey Davy in 1815. It will indicate the presence of *fire-damp* in a mine, which, when mixed with certain proportions of atmospheric air, becomes ignited within the gauze cylinder forming the "top," or upper part of the lamp. The flame, however, cannot pass through the wire gauze and set fire to the gas outside. There is no glass used in the construction of this lamp; it consists simply of a brass cistern for the oil, with wick, &c., surmounted by a chimney or cap of iron, or copper wire gauze, having not less than 784 (28 × 28) apertures to the square inch. Diameter of gauze is about 1½ inch, and about 8 ins. in height. The *Davy* is not a safe lamp to work with under certain conditions.

DAY (Pa.). The entrance to a mine on a hill-side.

DAY-EYES (N. W.). Inclined planes driven from the surface to *win* and *get* the mines.

DAY-HOLE. Any heading, or level from the surface communicating with the mine.

DAY-MEN (Y.). Men employed in building *packs*, and performing other work in the mine, for which they are paid by the day, or by *time*.

DAY-SHIFT. When a colliery is worked by two *shifts*, or relays of men, that which works during the daytime is called the *day-shift*.

D. C. *Downcast Shaft.* See *Downcast*.

DEAD. 1. An unventilated or airless *heading* or *working*.

2. The *creep* after subsidence or upheaval has taken place to the full extent.

DEAD GROUND. A faulty or barren piece or area of coal strata.

DEADING (G., Som.). See *Deadwork*.

DEAD RENT. A certain, fixed, or *minimum rent* paid at specified times by a lessee of a mine, whether minerals are worked and sold or not.

DEAD-SMALL (N.). The smallest coal which passes through the screening or separating apparatus, being almost as fine as dust.

DEAD-WORK. The work of driving out into a mine

for the purpose of proving and preparing to work it, or work which at the time produces little or no profit.

DECK. The platform or level upon which the tubs and men *ride* on a *cage*. Cages are occasionally made with as many as four *decks*.

DECKING. The operation of changing the *tubs* on a *cage* at top and bottom of a *shaft*. There are several very ingenious contrivances for performing this by mechanical means. One is Fowler's hydraulic loading and unloading apparatus, whereby each deck is operated upon simultaneously. The loaded tubs are at some collieries withdrawn from the cages by steam power, whilst the empties run into them by gravity. See *Onsetting Machine*.

DEEDS (N.). Débris of pit refuse tipped upon the *spoilbank*.

DEEP. Workings below the level of the pit bottom or main levels extending therefrom.

DEEP COAL. *Coal seams* lying at a depth below the surface of over, say, 600 or 700 yards.

DEEP PIT. A pit-shaft exceeding 400 or 500 yards in depth.

DELF (F.D., L.). A *vein, seam, mine*, or *bed* of *coal* or *ironstone*.

DEPUTY. 1. (N.) A man who fixes and withdraws the timber supporting the roof of a mine, and who attends to the safety of the *roof* and *sides*, builds *stoppings*, puts up *bratticing*, and looks after the safety of the *hewers*, &c., generally one deputy to every 12 workmen.

2. (M.) An underground official who sees to the

general safety of a certain number of *stalls* or of a *district*, but who does not set the timber himself although he has to see that it is properly and sufficiently done. He will often have the overlooking of as many as 100 men and boys.

DEPUTY SYSTEM (N.). The plan of having all the timbering or propping of the working places performed by *deputies* (1) specially appointed.

DERRICK. A high frame or *head gear* constructed of timber poles, placed over a *bore-hole* (1), upon which is fixed or hung a pulley or sheaf for carrying the rope by which the *rods* (2) are lifted.

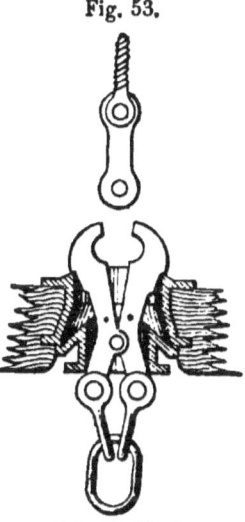

Fig. 53.

DETACHING HOOK. A self-acting mechanical contrivance for setting free a *winding rope* from a *cage*, &c., when the latter is raised beyond a certain point in the *head gear;* the rope being released, the cage remains suspended in the frame.

There have been a number invented, and a variety of them are in use. Fig. 53 is a sketch showing the action of one which has been much used.

Walker's Hook.

DEVIL. A *back-stay*, also a kind of *jockey*.

DIAGONAL STAPLE (N.). A shallow pit or shaft sunk in a sloping or diagonal direction at the back end of the main beam of a pumping engine in which the lever-beam works, so that the work of pumping may be divided between the two ends of the main beam.

DIAL. 1. A circumferentor or compass fitted with sights, spirit levels, and vernier, for making underground surveys.

2. To survey with a *dial* (1) and chain. See *Dialling*.

DIALLING. The operation of making a survey with the *dial*. There are two ways of using the instrument known as *loose needle* and *fast needle* dialling. The former is practised when all the angles or bearings of the different roads are taken (when such roads are free from iron tram-rails, &c., which attract the needle of the dial and give erroneous readings), by "*reading the needle*," as it is called. In the latter method the needle is only consulted in the first *sight* or at the commencement of the survey (all iron being removed from near the instrument), all subsequent angles being read off from the vernier, so that the presence of iron has no effect upon the work. See *Latch*.

DIAMOND CHISEL. A cutting chisel used in boring for coal, &c., having a diamond or V shaped point.

DIAMOND SYSTEM. Boring for coal, &c., with diamonds or carbonates, which are stones of a coarse quality and of a black colour. In this system the rock is cut or removed by abrasion, the boring rods or rather tubes, for they are hollow, are caused to revolve or rotate very rapidly (there being no percussive action whatever) up to 250 revolutions per minute. Entire *cores* are secured whereby the precise character of the various beds bored through are determined. The *débris* or *bore meal* is removed from the hole, as fast as it is made, by the constant flow of a stream of water forced down inside the rods and carrying up the stuff to the surface. The work is performed by steam machinery, and a very rapid progress is often made, say 10 feet per day as an average for a hole 1000 feet deep; but of course everything depends upon the nature of the strata bored through and the care bestowed upon the working of the machinery.

DIBHOLE (L.). The lowest part of a *pit shaft* below the scaffold on which the *cages* drop. It forms a water *lodge* for the drainage of the mine, out of which it is raised to the surface. See *Sump*.

DICE (Lei.). The layers in a coal seam of a glossy bituminous nature which naturally break or split up into small square pieces resembling *dice* in shape.

DIFFERENTIAL PUMPING ENGINE. A compound direct-acting pumping-engine, generally of the horizontal class, and usually fixed at the pit bottom for forcing the water direct to surface. So called, because it is fitted with differential valve gear of a very effective and ingenious type, the invention of a Mr. Davey of Leeds.

DILLY (N.). A counter-balance mounted upon two pairs of tram wheels by means of which the empty tubs are carried up an underground incline of a greater inclination than 1 in 3.

DILSH (S. W.). Inferior *culm* in the shape of a thin stratum.

DIP. 1. To slope downwards from the surface.

2. A heading or other underground *way* driven to the *deep*.

3. Inclination of strata when viewed in the direction

of the fall. The amount of *dip* is said to be 1 in so much, e.g. 1 in 4. Or, so many inches in the yard (9" in the yard), or, in degrees (14°).

4. (N. S.) A *heading* driven to the full *rise* in steep mines. It is usual to drive a *pair of dips* about 10 yards apart every 180 yards or so, out of the levels which run at right angles to the *cruts*, and out of these *dips* are driven *cross headings* right and left on the *strike*, about 10 yards apart, commencing at the upper end first and working downwards (see *Drifting Back*) Fig. 54.

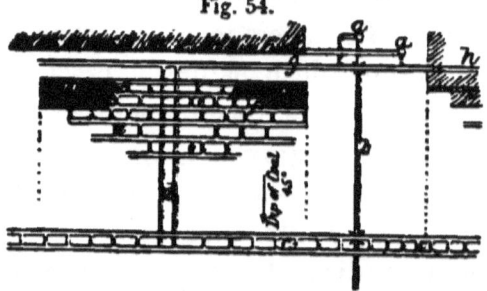

Fig. 54.

*a*, Shafts. *b*, Crut. *c*, Levels in coal. *d*, Dips (pair of) rising 1 in 1. *e*, Cross headings. *f*, Face of drifting back. *g*, Return airway. *h*, Goaf.

DIP JOINTS (Pa.). See *Backs*.

DIPPER (N.). A *down-throw, fault*, which see.

DIPPING (S. W.). A *dip* (2).

DIPPLE. See *Dip* (2).

DIP SPLIT. A current of *intake* air directed into or down a *dip* or *deep district* of a mine.

DINT (M.). See *Bate*.

DIRT. 1. Clay, bind, or other useless rubbish produced in mining, and which accidentally is sent out of the pit mixed with the coal.

2. (N.)  Foul air or *fire damp*.

DIRT BED or BAND. A thin stratum of soft earthy refuse interbedded with coal seams.

DISH (N.). The length or portion of an underground engine plane nearest to the pit bottom, upon which the empty *set* stands before being drawn *inbye*.

DISLOCATION. A fault of fracture of the strata as shown in Fig. 60.

DISTANCE BLOCKS. Pitch pine blocks placed in between the main spears and the side pump-rods by which the proper distance between them is adjusted. See Fig. 55.

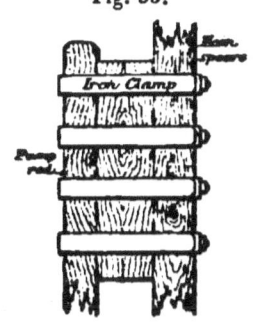

Fig. 55.

DISTRICT. A limited area of underground workings. Collieries are usually divided into several districts. As far as is possible each should be provided with a separate *split* of fresh air and a distinct *return* air-way leading to the *upcast shaft*. There is generally a *deputy* (2) or *overman* for every *district*.

DITCH (Lei.). To go stiff. To clog. To impede.

DITCHED TOP (Lei.). A coal-seam which has a hard unyielding *top*, and is with difficulty separated from the *roof*, is said to have a *ditched top*.

DOBBY WAGON (Y.). A cart into which dirt out of the mine is tipped.

DO (doo) (Lei. D.). See *Bout*.

DOCK (N.).

Dog. An iron bar, spiked at the ends, with which timbers are held together or steadied.

Dog and Chain. An iron lever with a chain attached by which *props* are withdrawn from the *goaf.* Fig. 56 is a sketch showing the way in which a dog and chain is used.

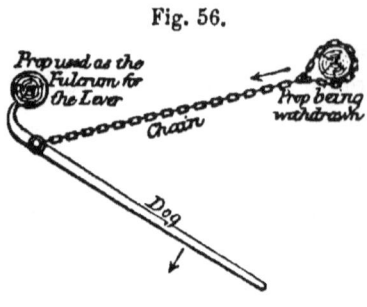

Fig. 56.

Dog-belt (M.). A strong broad piece of leather buckled round the waist, to which a short piece of chain is attached, passing between the legs of the man or boy drawing a *dan* (2) in the workings.

Dogger (Cl.). A bed of inferior *ironstone* overlying the main *seam.*

Doggy (S. S., N.). An overlooker of a certain number of boys and men in a pit. See *Corporal.*

Dogs (Som.). See *Cage Shutes*, but generally made longer than in Fig. 35.

Dolly (S. S.). A cast-iron weight suspended over the men when riding in the *shaft*, to act as a counterbalance to the *winding engine.*

Domed. Dipping away in all directions from a centre.

Dook (S.). An underground inclined plane to the *deep.*

Doors. Wooden doors, either single or double, fixed in underground roads of all descriptions to serve as *stoppings.* They are always fixed so as only to open towards the *intake* air. Every door in a pit should be so hung and otherwise adjusted that it will close of itself.

DOUBLE-BANK CAGES (S. W.). Cages having two *decks*, or a multiple of two, so that *decking* may be performed at two levels or *banks*.

DOUBLE CRIB. Two *wedging cribs* placed one on the top of another.

DOUBLES (Som.). The repeated folds or overlaps of the coal strata in the Radstock district. Fig. 57 is a section of a coal seam exhibiting *doubles* in a very marked manner.

Fig. 57.

DOUBLE SHIFT. A colliery is said to be working *double shift* when there are two *shifts* of *colliers* (1) employed in *getting* coal.

DOUBLE STALL (S. W.). A system of working coal

Fig. 58.

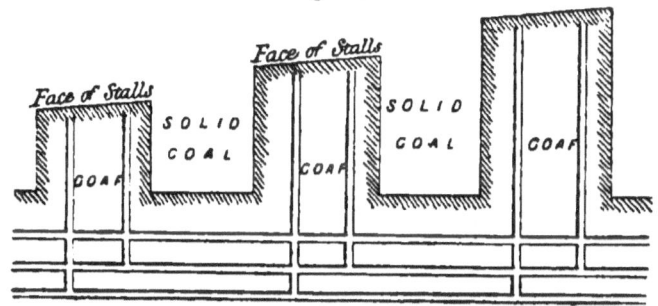

in which the roof falls within chambers or *banks* (4) of a limited width. See plan, Fig. 58.

G 2

**Double Timber** (S. W.). Two props and a *bar* placed across the tops of them, in the form shown in Fig. 59, for giving support to the *roof* and sides of a *heading* or *way*.

Fig. 59.

**Double Working** (N.). Two *hewers* working together in the same *heading*.

**Douce.** To beat out or extinguish an accidentally ignited jet of *firedamp*.

**Down.** Underground. In the pit.

**Down Brow** (L.). A *dip incline* underground.

**Down-cast.** 1. The *shaft* through which the *intake*, or fresh air, enters a mine, and the one used for winding coals in, and in which the pumps are generally fixed. It is usually circular in form, though sometimes rectangular and oval. Shafts are now sunk up to 18 and 20 ft. in diameter within the *walling*. The deepest in Great Britain is 939 yards (Ashton Moss, near Manchester). See *Signs*.

2. A *fault* which throws a coal-seam downwards. See *Down-leap*.

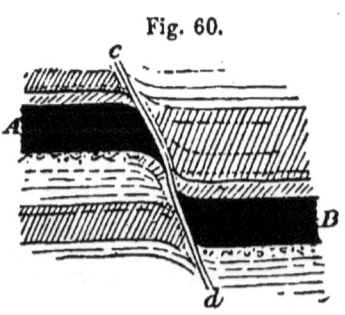

Fig. 60.

**Downer** (Som.). A rest or cessation from work, say half an hour taken during a *shift* or *turn* (1).

**Down-leap** (M.). A dislocation of strata which has caused a coal seam to be abruptly cut off and be brought below its original level.

In going from A to B in Fig. 60 the line $c\ d$ will represent a *down-leap*.

Down Spouts (L.). Pipes fixed down the sides of a *shaft* for conducting water from one *garland* (1) to another.

Down-throw. See *Down-leap*.

Dowzing Rod (Som.). The *virgula divinitoria* or divining-rod. Formerly commonly used in attempting to discover minerals. It consisted of a forked branch off a hazel tree in the form of a Y. One end of the rod was supposed to point in the direction of the mine when carried in a particular way over the ground to be examined. The person carrying the stick was called the *dowzer*, and the practice of using it was known as *dowzing*. A remnant of ancient superstition.

Draft (S. W.). *Allowance coal*. About 360 lbs. per week to every householder.

Drag. 1. The frictional resistance produced by the current of air circulating in a mine, the amount of which depends upon the extent of *rubbing surface* as it is called—i.e. the length × the perimeter—of the air ways. The *ventilating pressure* necessary to overcome the *drag* increases and decreases in proportion as the extent of rubbing surface increases or decreases, and varies in proportion as the square of the velocity of the air current increases or decreases. Therefore in order to double the quantity of air passing through an air-way the power to produce it would have to be increased fourfold, because there would be a fourfold resistance in the shape of friction (drag) to be overcome. In the same way half as much air would only take one quarter the pressure.

2. See *Back-stay*.

3. A scotch (either a short wooden or an iron bar) placed between the spokes of the wheels of trams to check their speed upon an inclined way.

DRAGON (S. S.). A kind of barrel in which water is raised from a *gin pit*.

DRAGS-MAN (N.). A man employed as a *putter* or pusher of *tubs* about underground in the working places.

DRAG-TWIST. A scraper with a spiral hook at one end with which the *bore meal* is extracted from a bore hole.

DRAW (S. S.). Strictly speaking, the distance on the surface to which the subsidence or *creep* extends beyond the *workings*. See *Creeping*.

DRAWER (S.). One who pushes *trams* underground, or drives a horse or pony drawing minerals to the pit bottom, or on to an engine plane or *jig*.

DRAWING. 1. Recovering the *prop wood, chocks*, &c., from the *goaves* for using over again. This work is commonly performed with the use of the *Dog and Chain*, which see.

2. Knocking away the *sprags* from beneath the coal after *holing*.

3. Raising coal, &c., up a *pit shaft*, or up a slope or inclined plane.

DRAWING A JUD (N.). Bringing down the *face* of coal, previously set free to fall by withdrawing the *sprags* after *kirving*.

DRAWING ENGINE. The engine by which the minerals are raised from the mine, by which the men and materials are lowered and raised, and by which the water produced in the workings is sometimes raised either by pumps worked from the same engine, or in tanks or barrels attached to the *winding rope* or riding in the cages. See also *Winding Engine.*

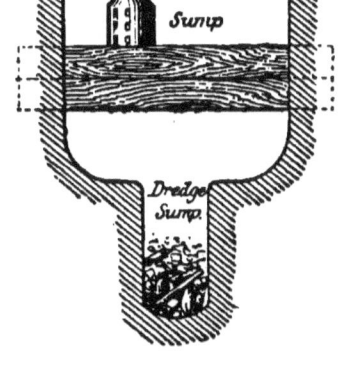

Fig. 61.

DRAW SMALL. When a *winding rope*, from the effects of wear and tear, has become less in diameter or in thickness from that cause, it is said to be *drawing small.*

DREDGE SUMP (N.). A small reservoir at the bottom of a pumping shaft, in which the water collects and deposits any sediment or débris, and is pumped up clear. Fig. 61.

DRESSANTS (F.). *Rearers* or very steep lying seams of coal, &c.

DRESSER (M.). A tool used by colliers and banksmen for splitting up large lumps of coal, and for dressing off *dirt* or *brasses* when cleaning coals for the market. See Fig. 62.

Fig. 62.

DRESSING (M.). Trimming and cleaning up a *stall face* after the *loaders* have left off work, and before the *holers* commence work. This work is performed at night.

DRIFT. 1. An underground gallery driven across or obliquely to the planes of stratification. See *Branch*.

2. An inclined plane driven entirely in a coal seam. The work of making a drift is known in mining language as *drifting*.

3. (F. D.) A hard shale. ·

4. (N.) A *head* (1) driven on the *strike* of the *coal seam*.

DRIFT AND PILLAR (N. S.). A system of working coal not unlike the *bankwork* of Yorkshire.

DRIFTING BACK (N. S.). The operation of working away the pillars towards the pit bottom in *rearers*. *Drifting back* commences as soon as the *cross headings* are driven out.

DRIFTING CURB. An oak *curb* forced downwards through quicksand, having a circle of planks driven down all round at the back of it to keep out the sand and water.

DRILLING (U. S. A.). Boring deep holes in search of coal.

DRIVE. To excavate horizontally, or at an inclination, places not more than a few yards in width underground.

DRIVERS (M.). Men who break down the coal in the *stalls* with hammers and wedges, after the *holing* is finished.

DRIVING. 1. A long narrow underground excavation or *heading* (1).

2. (B.) A *stone head* (1) driven through a *fault*, &c.

DRIVING BY LINES. Keeping the axis of the *heading* being driven exactly true to a certain bearing or degree of the *dial*. Two lines, or strings, steadied by weights are suspended from *stomps* fixed in the *roof* from three to six feet apart; the prolongation of the line drawn between them being the bearing or proper direction, or *point* as it is commonly called, of the heading.

DROP. 1. To lower coals down from a higher to a lower level on the pit bank, or at pit bottom, when the *decking* is performed in one operation, or when the *cage* is only moved once during *decking*.

2. (N.) A shoot down which coals are run into *keels* or boats.

3. To allow the upper *lift* of a seam of coal, &c., being worked, to fall or *drop* down, when the lower portion is first *gotten*. See Fig. 63.

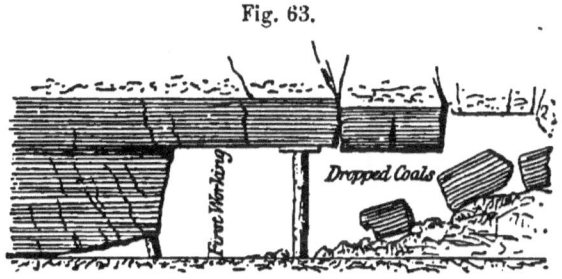

Fig. 63.

4. A general reduction of wages in the coal trade.

DROP PIT. A shallow *pit shaft* in a mine, in which coals are lowered in *tubs* upon *cages* by means of a *clip pulley*, or brake-wheel, from one seam to another, or, where a *fault* exists, from the higher to the lower level. The principle upon which it is worked is similar to that of a self-acting inclined plane, viz. the weight of the coals dropped being greater than that of the rope, and friction of the empty tub and appliances.

DROP SHEETS (N.). Doors made of canvas, by which the ventilating current is directed and regulated through the workings.

DROSS (S.). Very small coal-dust, or slack.

DROSSY COAL (D.). Coal with iron pyrites.

DROWNED-OUT. Flooded. Mines under water.

DROWNED WASTE. Old workings full of water.

DRUB (Y.).

DRUM. 1. That part of the *winding engines* upon which the *winding-ropes* are coiled or wound. They are constructed in various forms (see end views or plans, Fig. 64), of diameters ranging from 5 to 32 feet, according to depth of shafts and size of ropes, &c. See

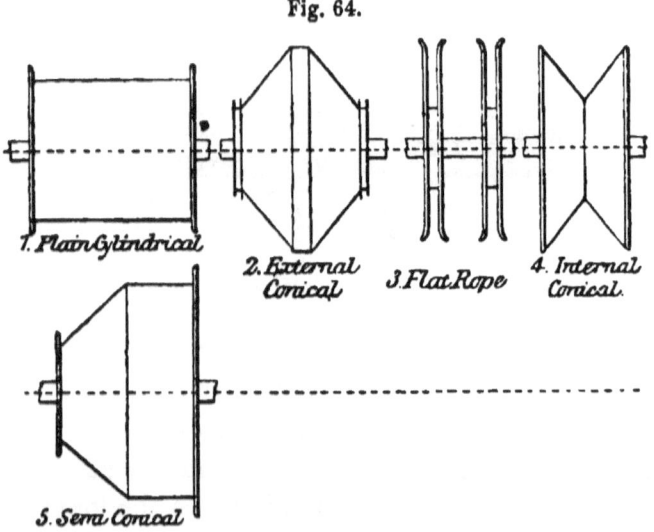

Fig. 64.

1. Plain Cylindrical.  2. External Conical.  3. Flat Rope.  4. Internal Conical.

5. Semi Conical.

*Conical Drum.* The usual number of revolutions made per run is from 20 to 30.

2. The barrel or roll upon which a self-acting incline

rope is coiled, generally made in the form of No. 1, Fig. 64.

3. (L.) A brick, iron, or wooden cylinder, with which beds of sand are sunk through. See *Running the Drum.*

DRUM-HEAD (N.). A short heading formed to the rise of a level, or *bank-head*, in which the *drum* of a self-acting inclined plane is fixed.

DRUM-HORNS. Wrought-iron arms or spokes projecting beyond the surface or periphery of flat-rope drums, between which the ropes coil or lap, the tips being often connected by a ring of iron riveted on.

DRUM-PULLEY. A pulley-wheel used in place of a *drum* (1). See *Köepe System*, Fig. 89.

DRUM-RINGS. Cast iron wheels, with projections, to which are bolted the staves or laggings forming the surface for the ropes to lap upon. The outside rings are shrouded, to prevent the ropes from slipping off the sides of the drum.

DRY (S.). A joint in the *roof* of a coal-seam, which cannot usually be discovered until the roof falls. They frequently exist in connection with *lypes*.

DRY COAL. That which contains but little hydrogen. For instance, the "Aberdare 4 Feet" seam of Glamorganshire, a first-class *steam coal.*

DRY SEPARATION. The systems upon which coal is screened and further separated by taking out the small pieces of shale, pyrites, &c. (*dirt*, 1), by what is called the *wind method*, i. e. the force of a blast of air is directed upon the screened coal, and thereby separates it into various sizes due to their specific gravity. See *Wind Method.*

DUAL-ROPE (Y.). A hemp capstan rope upon which men ride in an *engine-pit*.

DUFF. See *Dross*.

DUKEY (Som.). 1. A large carriage or platform running upon wheels on rails working on a *dip* inclined plane underground, upon which a number of small *trams* of coal are raised by engine-power at one operation. So named after the double coach called the "Duke of Beaufort."

2. (S. W.) An inclined plane worked by engine-power.

DUKE-WAY (Som.). The plan of drawing coals up a dip incline to the pit-bottom by a rope worked by the *winding-engine* at surface, the other rope working the *cage* in the shaft simultaneously, i. e. whilst the cage is going up, the empty trams are running down the incline, and *vice versa*.

DUKEY-RIDER (S. W.). A boy who accompanies the *train* of *trams* running upon a *dukey* (2).

DULL (B.). Slack *ventilation*. Insufficient *air* in a pit.

DUMB DRIFT. A short tunnel or passage connecting the main *return airways* of a mine with the bottom of the *up-cast* shaft, in order to prevent the return air from passing through and over the ventilating *furnace*.

DUMB FURNACE. See *Dumb Drift* and *Cold Furnace*.

DUMP (Pa.). To throw coals, &c., by tilting up the *car* into, or shooting them down a dip road in a pit, or upon the inclined plane of a *breaker* to a loading stage.

DUMMY (N. S.). A low truck on four wheels running

upon rails, and loaded with pig iron or some other heavy material; employed in steep seams or *rearers* as a balance-weight to bring up an empty *tub* (1) on an inclined plane or a *dip* (4); the weight of the coals, &c., in the tub being sufficient to overcome the resistance of the *dummy* when being *braked* down.

DUNN BASS (L.). A description of *Bass*.

DUNS (G.). Argillaceous shale. See *Cliff*.

DUNSTONE. 1. (D.) Ironstone in beds or seams.
2. (S. W.) Hard kind of *fire-clay*, or *under-clay*.

DUN-WHIN (N.). A rock commonly met with in the *coal measures*.

DUST. 1. Fine black powdery substance adhering to the timbers, &c., in a coal mine. See *Coal Dust*.
2. See *Dross*.

DUSTERS (S. W.). Men employed in cleaning *trams* of dust and dirt in and about mines.

DUST EXPLOSION. An explosion of coal-dust mixed with a small percentage of *fire-damp*.

DUTY (of a Cornish pumping engine). The number of pounds weight of water raised one foot high with a consumption of 112 lbs. of coal.

DYKE or DIKE. An intrusive band or vein of hard rock, usually of igneous origin. In the north of England a *fault* is often called a *dyke*. They are not always accompanied by a dislocation of the strata— probably have their origin in some deep-seated connection with the molten interior of the earth, out of which they have doubtless been ejected in the shape of lava, at a period subsequent to the deposition of the *coal measures*—extend in almost straight lines through the country, in one case upwards of 70 miles. Though generally taking a vertical line, like a wall, frequently are discovered lying at different angles, and even interbedded with seams of coal, &c., and in almost all cases when in proximity to a *trap dyke*, the coal and other rocks are partially coked and calcined from the heat of the lava when first injected into the fissures it occupies.

# E.

EARS (D.). Small iron loops or rings fixed on the sides of *tubs*, &c., to which *side-chains* are attached.

EARTH. A term used for soft shaly or clayey ground met with in *sinking* through the *coal measures*.

EARTH COAL. A name sometimes given to Lignite—earthy brown coal.

EAT OUT (N.). To turn a *heading* or *holing* to one side in order to *win* the coal on the other side of a *fault* without altering the level course of the heading. In Fig. 65 is given a plan and section showing two cases of *eating out* a fault. The side to which the heading must be driven on meeting with the fault *a b* depends entirely upon two things—the nature of the fault (whether an *up-throw* or a *down-throw*), and the *dip* of the coal on the far side of it. In No. 1 case the fault is a *down-throw*, coal dipping to the right; and in No. 2 the fault is *up*, and the dip to the left; and so, in order to *win* the coal beyond *a b*, the *eating-out* must be done in both cases on the *left*. Had, however, the

USED IN COAL MINING, ETC. 95

dip in No. 2 been reversed, the *eating-out* heading must have been on the right at or about C. The fault being of 4 yards *throw*, and the dip 1 in 4, it follows that the

Fig. 65.

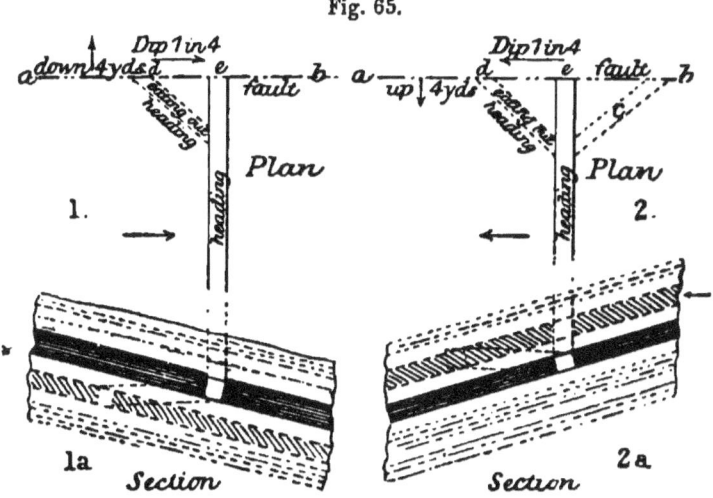

distance to be followed alongside the fault before meeting with the coal again, or from *d* to *e*, will be 16 yards.

EDGE COALS, EDGE METALS, EDGE SEAMS .(S.). Highly inclined seams of coal, or those having a dip greater than say 30 degrees. See Fig. 66.

Fig. 66.

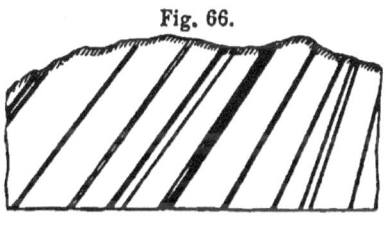

EGG COAL (Pa.). *Anthracite* which passes over a 2½ inch *screen*.

EMPTIES. Empty *trams*.

EMPTY ROPE. Any *winding* or *hauling* rope from which the load upon it has been removed.

END. The inner extremity of a *head* (1) or *stall*.

END or END-ON. Working a seam of coal, &c., at right angles to the *cleat*, or natural planes of cleavage.

ENDING (M.). See *Bolthole*.

ENDLESS CHAIN. A system of underground *haulage*, (used also on the surface) in which the *trams* are drawn along the *ways* by a chain worked by an engine from and to the *shafts* to the branch *roads* or *gates* leading to the *working places*. They are attached separately to the main chain at intervals of from 10 to 30 yards; the speed of the chain being about three miles an hour. Applicable to mines not having much inclination.

ENDLESS ROPE. 1. A system of *haulage* carried out and arranged in much the same way as the *endless chain*, and especially applicable to seams having a moderate inclination. The *trams* are attached to the rope either singly, in pairs, or in sets of 30 or 40, and the speed is slow. For different ways of attaching trams to endless chains and ropes see *Haulage Clip*. Fig. 67 is a plan showing the *endless rope* system as applied

Fig. 67.

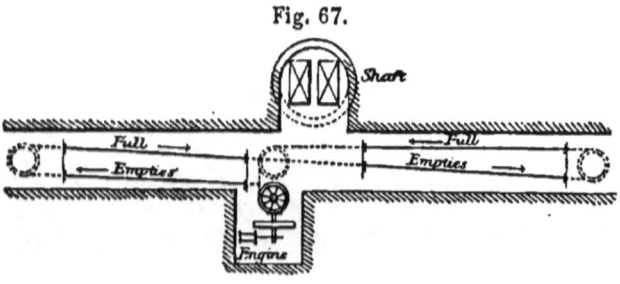

to moving the trams about in the vicinity of a shaft bottom.

2. A new system of *winding*, in which the rope

passes through the *cages* being secured beneath them by wrought iron clamps, by shifting which the distance between each *cage* can be altered at will, thus making it possible to hoist at different times from different levels without losing the advantage and economy of balanced cages. The endless rope runs in a deeply grooved pulley driven by a pair of engines.

ENDS (Y.). *Headings* which are driven on the *end* or *end-on*.

ENGINE. A collier's term for engine-house or building, *arching*, &c., within which a steam-engine is fixed.

ENGINEER. 1. (N.) The person at a colliery having charge of the whole of the machinery both on surface and underground, and of the workshops.

2. (S. W.) The *brakesman* or engine-man.

3. (M.) The mining engineer or *viewer*.

ENGINE PIT. A *shaft* used entirely for pumping purposes.

ENGINE-KEEPER (S.). See *Brakesman*.

ENGINE-MAN. One who works a winding, hauling, fan, pumping or other engine.

ENGINE PLANE. An underground *way* either level or dipping *inbye* or *outbye* or both (undulating) along which the *tubs* are conveyed to and from the *workings* to the pit bottom by engine power. See *Endless Chain, Endless Rope, Main Rope, Tail Rope*.

ENGINE TENTER (N. S.). See *Brakesman*.

ENGINEWRIGHT (M.). A thoroughly practical man, whose duty about a colliery is to daily inspect the external parts of the machinery, ropes, and other

H

appliances, and to see that the same are kept in efficient working order—who has the control of the smiths, and other surface workmen, and takes the leading part in superintending the erection or fitting up of most of the machinery and other matters connected with the mechanical engineering of collieries.

ESCAPE. A second or additional *shaft* by which the men are got out of the mine in case of accident to the other shafts. Also an *upcast*.

ÉTAGES (F.). See *Face, Mouthing, Level*.

ETTLE (N.). See *Attle*.

EVERLASTING LAMPS (N.). Natural jets of *fire-damp* or small *blowers* set fire to and continuing to burn as long as *gas* was given off. One of these *lamps* is said to have been burning for 19 years in the Newcastle coal field. The gas was conveyed to the surface in pipes and there set fire to.

EXPLOSION. The sudden ignition of a body of *fire-damp* in a mine (often aggravated by an admixture of *coal dust*), so often carrying death and destruction all before it. The fearful blowing up of the Oaks Colliery in South Yorkshire, on the 12th December, 1866, when 371 men and lads were lost, is the most disastrous one which has ever taken place.

There appear also to be two other causes of *explosions* in coal mines, though fortunately probably seldom taking place, viz. 1. The ignition of inflammable gases evolved from a *standing fire* or burning or mouldering coal. 2. The sudden ignition of bisulphuret of carbon, which is given off by coal and explodes at a very low temperature, even in the absence of flame.

EXTINCTEUR (F.). A machine of rather recent invention which discharges on to a burning mass of coal, water charged with carbonic acid under a very high pressure—a sort of soda-water. A man carries the apparatus on his back and projects the gaseous water by means of a hose like that of a fire-engine.

EYE (Y.). The mouth or top of a *pit-shaft*.

## F.

FACE. 1. The place at which the coal is actually being worked away either in a *stall* or in a *heading*.

2. A *cleat* or *back*.

3. (L.) To place a full *tub* in position for being lowered down a *brow* or *jig*.

FACE AIRING (N.). That system of ventilating the workings which excludes the airing of the *goaves*; that is to say, nearly the whole of the air is made to sweep through the pit, ventilating the working *faces* and main roads only.

FACE ON. The reverse of *end on*, or working a mine parallel to the *cleat* or *face* (2). In order to extract the coal in the largest possible lumps it will generally be found advisable to keep the *face* line of the *stall* neither fully *face on* nor *end on*, but say half-and-half, or any other convenient angle. See *Horn Coal*.

FACING. See *Cleat*.

FAHRKUNST (Belg.). An apparatus for lowering and raising the colliers, &c., in a shaft. See *Man Engine*.

FAIKS (N. and S.). Shaley and slatey strata more or less gritty.

FAIRING (C.). Kindly treating pit ponies by boys.

FAKE. See *Faiks*.

FALL. 1. A mass of *roof* or *side* which has fallen in in any subterranean working or gallery, resulting from any cause whatever. Immense *falls* take place generally immediately after a heavy *explosion* of *fire-damp*.

2. To blast or wedge down coal, &c., in the process of working it.

3. A length of *face* undergoing *holing* or breaking down for loading up.

4. To crumble or break up small from exposure to the weather; clays, shales, &c., *fall*.

FALLERS (L.). See *Cage Shuts*.

FALLING (N.). Thin shaley beds of stone, &c., taken down with the coal, above which a good roof may be met with.

FALLS (F.). Working by Falls. A system of working a thick seam of coal by falling or breaking down the upper part after the lower portion has been gotten.

FAN. A centrifugal mechanical *ventilator* driven by steam power. They are made up to about 46 feet in diameter. Several kinds are in use, the Guibal, Rammel, Waddle, Schiele, and others; some of them being able to produce a ventilation, under favourable conditions, of between 200,000 and 300,000 cubic feet per minute. The principle of the *fan* is that exhaustion or suction of the air out of the mine is produced by the rapid revolution of the blades of the machine, whereby a partial vacuum is created, and the air from the mine rushes in to fill it. Sometimes two fans are placed side by side and both kept running, or one in reserve in case of accident. The engine also to drive a fan is generally in duplicate. See *Ventilator*.

FAN DRIFT. A short tunnel leading from a short distance from the top of the upcast shaft to the *fan* chamber or casing in which the *fan* runs, along which the whole of the *return* air is drawn by the *fan*. In it, opening upwards, are occasionally fixed some wooden doors, intended to blow open in the event of a serious *explosion* taking place, and so save the fan from becoming seriously damaged.

FANGING (M.). *Bratticing* much the same in form as *trumpeting*, which see.

FANNERS (S.). A kind of rude form of *blow-george*.

FANS, and sometimes FANGS (S.W.). See *Cage Shuts*.

FAN-SHAFT. 1. A shallow *pit-shaft* sunk beneath a *fan* connecting it with the *fan drift*.

2. The *upcast shaft* where a *fan* is in use.

FARE (S.W.). Standing coal, or coal unholed or uncut.

FAREWELL ROCK. The *Millstone Grit*, embracing a series of strata unproductive in coal, and in which conglomerate and coarse siliceous grits often preponderate.

FAR-SET (M.). To timber up and *sprag* the far end of a *stall*, preparatory to *holing*.

FAST. 1. (L.) The first hard bed of rock met with after sinking through running sand or *quick* ground, upon which a *wedging crib* is generally laid.

2. When a *heading* or *board* end is not in communication with another one by a *bolt* or *thirl*, but has only one open end, it is said to be *fast* or called a *fast place*.

FAST BAY (L.).

FAST END. The limit of a *stall* in one direction, or where the *face* line of the adjoining *stall* is not *up* or level with, nor in advance of it. See Fig. 68. Three stalls are here shown; the *face* of the middle one is represented by the line *a b*; the end *a* is a *fast* end; that at *b* is called the *loose* end.

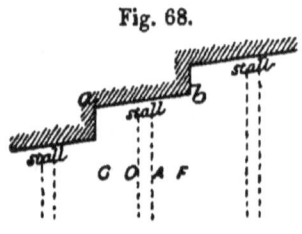

Fig. 68.

FAST NEEDLE. See *Dialling*.

FAST SHOT. A heavy or miss-shot. See *Shooting Fast*.

FAT COALS. Those which contain volatile oily matters; for example, the celebrated Cannel of Wigan.

FAULDING or FOLDING-BOARDS (S.). *Cage-catches* or *shuts* in *mid-workings*. Fig. 69 is a side elevation, showing the action of the catches.

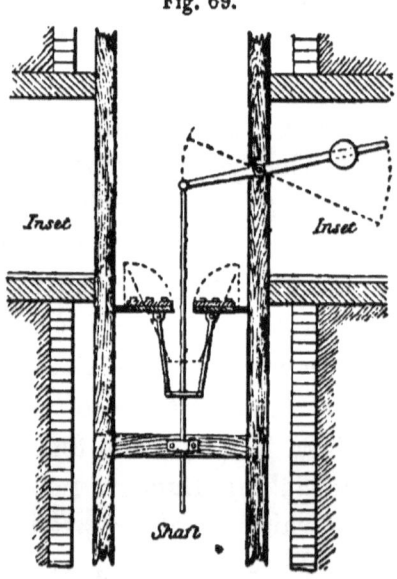

Fig. 69.

FAULT. Generally means a fracture or disturbance of the strata breaking the continuity of the beds. There are several kinds of *faults*, e. g. Faults of Dislocation, Fig. 70 (1); of Denudation (2); Upheaval (3); Trough

Fault (4); Reverse or Overlap Fault (5); Step Fault (6); Thinning out (7). Faults of displacement (1) are sometimes of many hundred yards *throw*, and run through the country for many miles. Those of type (2)

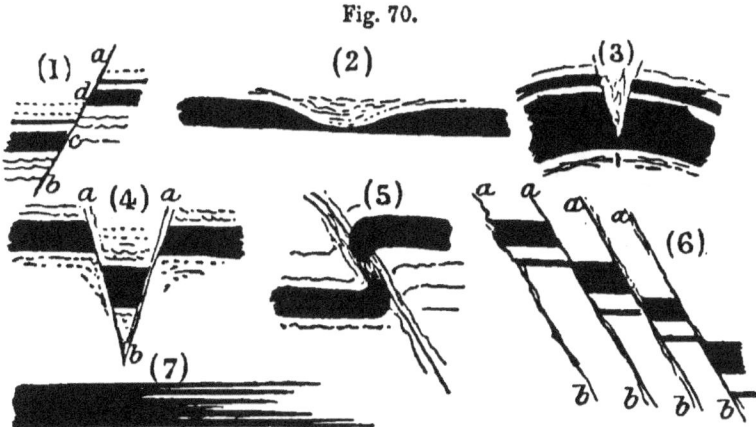

Fig. 70.

are frequently of great extent, being several hundred yards in width, and running through miles of country; (3), (4), and (5) are not of common occurrence; but (6) and (7) are types of faults met with in most *coalfields*.

FAULT-SLIP. The smooth surface of the fractured rocks at a *fault* of No. (1), (4), and (6) types, always to be found in the lines *a b*.

FEATHERS. Two long wedge-shaped pieces of steel or iron which are inserted at the back of a drill hole in coal, between which a long wedge is driven up, forcing the *feathers* apart, and thereby breaking down or loosening the coal.

FEE (M.). To load up the coal, &c., in a *heading* into *tubs*.

FEED. Forward motion imparted to the cutters or drills of rock-drilling or coal-cutting machinery, either hand or automatic.

FEEDER. 1. An underground spring or regular flow of water proceeding from the strata or from old coal or other workings.

2. A small *blower*.

FEER (M.). One who *fees*.

FEEL (S. S.). To examine the *roof* of a thick seam of coal with a long stick or rod by poking and knocking upon it.

FEIGH. Refuse coal or waste slack.

FENCE-GUARDS (S. S.). Rails fixed round the mouth of a *pit-shaft*, or across the shaft at an *inset* or at *mid-workings* to keep people and things from falling in.

FEND OFF BOB. A beam hinged at one end and

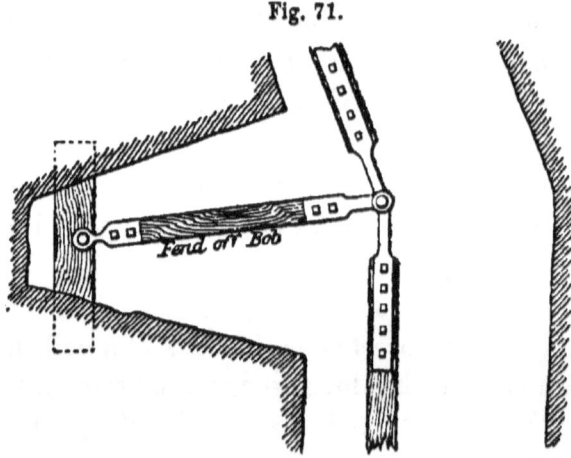

Fig. 71.

having a free reciprocating motion, fixed at a bend in a shaft or upon an inclined plane, to regulate the motion of and to guide the pump rods passing round the bend. See Fig. 71.

FETTLING (N.). Cleaning up and putting tidy any underground roadway, &c.

FIEG (S. W.). A crack in the roof, often letting in water.

FIELD. 1. A term used to signify a large tract or area of many square miles of coal. See *Coalfield*.

2. A *colliery*, or firm of colliery proprietors.

3. The immediate locality and surroundings of an *explosion*.

FIELD BOX (S. S.). A *colliery* accident club.

FIELD CLUB. A sick or accident club or society supported and managed by the Owners or Lessees of a *colliery*.

FIERY. Containing the explosive gas called *firedamp*, which see.

FIERY MINE. A colliery in which the seam or seams of coal being worked give off considerable quantities of light carburetted hydrogen gas. Mines subject to *blowers* are specially *fiery*. In England the mines of Lancashire, South Wales, Durham, and Yorkshire, are the most *fiery*.

FIGHTING. When the weight or pressure of the ventilating current of air in a mine becomes equal or nearly so in both the *downcast* and *upcast* shafts, and no appreciable movement is caused in the air, that is to say, when the motion of the air is first in one direction and then in another, the pit is said to be *fighting*.

FILL. To load *trams* in the mine.

FILLER. One who *fills* at a working place or in a *stall*.

FILLING. The places where *trams* are loaded in the *workings*.

FILTY (Som.). A local term for *fire-damp*.

FIND. A *sinking* or *driving* for coal, &c., attended with success.

FINGER GRIP. A tool used in boring for gripping the upper end of the *rods*.

FIRE. 1. A collier's term for the explosive *gas* met with in mines.

2. To blast with gunpowder.

3. To explode or blow up. The expression "the pit has *fired*" signifies that an explosion of *fire-damp* has taken place.

4. A *gob fire*.

5. A word painted upon a piece of board and fixed in the workings to indicate the presence of gas or other danger beyond it.

6. A word shouted out by colliers to warn one another when a *shot* is fired.

FIRE BANK (M.). A spoil-bank which takes fire spontaneously.

FIRE-BOSSES (U. S. A.). Underground officials who examine the mine for *gas*, and inspect every *safety lamp* taken into the colliery by the men.

FIRE-BOARD. A piece of board with the word *fire* painted upon it, and suspended to a *prop*, &c., in the *workings*, to caution men and lads not to take a naked light beyond it, or to pass it, without consent of the *underviewer* or his *deputies*.

FIRE BREEDING (S. S.). Any place underground showing indications of a *gob-fire*.

FIRE-CLAY. Any clay that will withstand a great heat without vitrifying. They contain from 60 per cent. to 95 per cent. of silica, and 2 per cent. to 30 per cent. alumina; lime or alkalies which act as a flux, being entirely absent.

FIRE-CUBE (S.). A rude kind of *furnace*, about 2 feet by 3 feet.

FIRE-DAMP. The explosive gas of coal mines. Light carburetted hydrogen. The chemical formula is $C_2H_4$. In every 100 parts of this gas there are generally 96 of *fire-damp*, 3·5 of nitrogen, and ·5 of carbonic acid gas. Being of very light specific gravity (air being 1 *fire-damp* = ·562 only), it is naturally always to be found in the highest points in the workings, that is to say, in the cavities of the *roof* in the *goaves*, &c. Unless mixed with four or five times its volume of air it will not take fire but extinguishes a light. It sometimes exists in the coal under the enormous pressure of 300 to 400 lbs. per square inch.

FIRE-ENGINE. A pump worked by hand for playing upon *gob-fires*.

"FIRE HEAVY." Words marked upon the scale of a mercurial barometer to indicate when much *fire-damp* may be expected to be given off in the mine, and to show that extra vigilance is required to keep the *ventilation* up to its full power.

FIRE-LAMP. 1. A rough description of iron basket on three legs or hung by chains from posts, in which coals are burnt to give light to *banksmen* where gas is not used. Fig. 72.

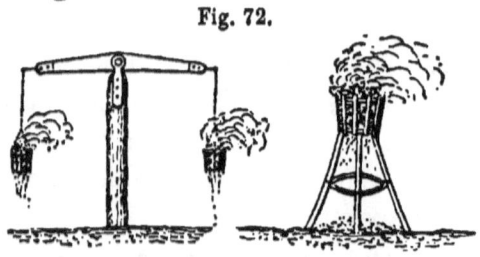

Fig. 72.

2. An iron bucket or basket of fire suspended in a pit-shaft (shallow mine) to create a draught or *ventilation* through the *workings*.

FIRE-MAN. A man whose duty it is to examine with a *safety lamp* the underground *workings* and *ways*, to ascertain if *gas* exist, to see to *doors, bratticing, stoppings*, &c., being in good order, and generally to ascertain that the *ventilation* of the mine is efficient.

FIRE-PAN (Y.). A kind of *fire-lamp* (2).

FIRE RIB (S. S.). A solid rib or wall of coal left unworked between *sides of work* to keep off *gob fires*.

FIRE-STINK. Smell, indicating spontaneous combustion in a coal pit.

FIRE-STONE (Som.). Synonymous with *Fire-clay*.

FIRE-TRIER (M.). See *Fireman*.

FIRING A MINE. Maliciously setting fire to a coal pit.

FIRING-LINE. A lighted candle attached to a string and drawn up over a long pole stuck in loose rubbish on the floor of the mine, until it came in contact with *fire-damp*, which was thereby exploded or *fired* in order to get rid of it. [A very objectionable way of clearing the workings of gas commonly practised in former days.]

FIRING-POINT. That at which *fire-damp* mixed with atmospheric air ignites or explodes. When there is four times as much air as gas the explosion is very feeble indeed, but increases in force as more air is added. 9 of air and 1 of gas causes the most violent explosion. When the proportion is 14 of air to 1 of gas the mixture ceases to ignite.

FIRST MAN (Lei.). The head *butty* or *coal getter* in a *stall*, who is appointed by the *manager* and is responsible for the safety of the men working under him, and for the proper working of the coal, which includes *holing, getting, filling, pack building, timbering, &c.* He maintains order and regularity amongst his fellow-workmen and in carrying on the work of *stalling*.

FIRST-WEIGHT. The first *weight* (2) which takes place after commencing to excavate any large area of coal, &c., without leaving *pillars*.

FIRST WORKING. *Winning* and proving a seam of coal, &c., by *heading* out into it and preparing to work the coal out by *longwall, banks, stalls, broken,* &c. *First working* is chiefly paid for by measurement, an allowance or *charter* being added, upon the tonnage. See *Second Working, Yardage.*

FISH-HEAD. An apparatus for withdrawing the *clacks* of pumps through the *column* (1).

FISSLE or FISTLE (N.). To make a faint crackling

noise, which takes place when *creep* begins in the *workings*.

FITTING (S.). The shafts and plant of a colliery.

FLAG (Ch.). A bed of hard marl stone overlying the *rock head* in salt mines.

FLAIKES (S.). Shaly or fissile sandstone.

FLAMPER (D.). *Clay ironstone* in beds or seams.

FLANCH (N.). The flange or broad ends of pump *trees* or other iron pipes where joined to one another.

FLANK HOLES. Holes bored into the sides of *headings* or other underground workings, to test the thickness of a rib or barrier, or the position of old workings likely or known to contain water or *gas*, or both.

FLANNELS. Suits of stout white flannel clothes provided by the *masters* for the *enginewright* and his assistant for wearing in an *engine-pit* or other wet place when doing repairs, &c.; also a flannel coat is often allowed to a *bottomer*, a *night watch*, &c.

FLAPPER-TOPPED AIR CROSSING. An *air crossing* fitted with a double door or valve giving direct communication between the two air currents when forced open by the *blast* of an *explosion*. The flappers or doors being so arranged that they should fall to or close of themselves immediately the blast is passed, and so restore the ventilation to its ordinary course. The object of the doors is to preserve the *overcast* from damage in the event of the pit *firing*.

FLAPS. Rectangular wooden valves about 24 inches × 18 inches × 1½ inch thick, hung vertically to the framework of the air chambers of the Nixon *Ventilator*. See *Ventilator*.

FLASH (Ch.). A subsidence of the surface due to the working of rock salt and pumping of brine.

FLAT. 1. (N.) A place underground at which *corves* are put upon the *rolleys*, or where *tubs* are run off and on into *cages*.

2. (D.) A *district* or set of *stalls* separated by *faults*, old *workings*, or barriers of solid coal.

FLAT COALS (S.). Seams of coal lying horizontal or at a low angle.

FLATMAN (N.). One who links the *tubs* together at the *flats* (1) or levels.

FLAT-NOSE SHELL. See *Cleanser*.

FLATS. 1. Subterraneous beds or sheets of trap rock or *whin*.

2. (N. S.) Tracts of coal-seams which lie at a moderate inclination in districts containing *rearers*.

FLAT SHEETS. Iron plates laid as a floor of the pit *bank* (2), upon which the coal *tubs* are easily moved about.

FLAT SHUTS (Y.). Heavy iron plates forming part of the *heapstead*.

FLATTING (D.). Drawing or *leading* coals underground with horses and lads.

FLEEK (M.). Coal or other rock is said to *fleek* off when humps or masses of it fall off from a *slip* or *fault* in the workings without giving warning, or without much labour in cutting, &c.

FLINT (Sh.). Fine grained sandstone suitable for building purposes.

FLITCHING (N. S.). Widening the sides of a *heading*.

FLOAT. A clean rent or fissure in strata unaccompanied by dislocation.

FLOOR. 1. The stratum of rock, &c., upon which a seam of coal, &c., immediately lies.

2. That part of any subterraneous gallery upon which you walk or upon which a tramway is laid.

FLOTZ. The German for *seam* or *bed*.

FLUE (S. W.). A *furnace*, which see.

FLUSH (M.). A small quantity of ignited *fire-damp*.

FLY-DOORS (N.). Doors in working roadways, opening either way.

FLYING REED (S. S). The thinning out or splitting up in a northerly direction of the "Thick coal" seam.

FOAL (N.). A small boy who assists a *putter*.

FOALEY BANT (D.). A cluster of three or four boys sitting in chain loops attached to a hemp rope a few feet above the heads of a bunch of several men (also riding in chains attached to the same rope) in which position they used formerly to ride up and down a *pit shaft*.

FOLLOWING DIRT (L.). Loose shale, &c., in a thin bed forming the *roof* of a coal seam, which has to be taken down in the *workings* in order to prevent it falling and thereby causing accidents.

FOLLOWING-IN. A *shift* arriving at a working place before the previous one has finished work.

FOLLOWING-UP BANK (Y.). A breadth of about 6 yards of coal taken off on either side of a *leading bank*.

FOOT. That part of the *face* of a *heading* next the *floor*.

FOOTRILL, FUTTERIL, and FOOTRAIL. The entrance to a mine by means of a level driven into a hill-side, or a *dip* road, up which coal is brought.

FOTHER (N.). A measure of coals, 17⅔ cwt., being an ordinary cartload for one horse.

FORCER. A pump by which the water is raised with a *ram* or plunger; in short, a force-pump.

FOUDROYAGE (F.). See *Falls*.

FOUL. A condition of the atmosphere of a mine, so mixed by any gases as to be unfit for respiration or working in.

FOUL COAL. *Faulty*, or otherwise unmarketable coal.

FOULS. Where seams of coal disappear for a certain space and are replaced by some foreign matter.

FOUND. When *sinking* or *driving* to find or prove a mine of coal, &c., as soon as it is met with it is said to have been *found*, or ascertained to *lie* and *be*.

FOUNDATION (M.). The shafts, machinery, buildings, railways, workshops, &c., of a colliery, commonly called a *plant*.

FOSSE (F. and Belg.). A colliery or coal pit.

FOSSIL (M.). A local term formerly used for a particular kind of rock bed met with in *sinking*. *Cank*, *lignite*, &c., were called by this name.

FRAME-DAM. A solid *stopping* or *dam* in a mine

I

constructed of timber balks in a watertight manner so as to entirely keep back and resist the pressure of a heavy head of water.

FRAME TUBBING. Solid wood *tubbing*, entirely composed of rings or *curbs* of wood about 8" × 6" square built up in segments and wedged to keep it watertight

FREE-DRAINAGE LEVEL. See *Adit*.

FREE MINER (F. D.). A man born within the hundred of St. Briavels, in the county of Gloucester, who has worked a year and a day in a mine.

FREE SHARE (Som.). A certain proportion of a *royalty* on coal, &c., paid to lessor by lessee.

FRENZIED (S. S.). Crushed by the *creep* or subsidence of the *cover*.

FUR. A deposit of lime and other minerals upon the sides of pumps, boilers, &c.

FURNACE. A large coal fire at or near to the bottom of an *upcast* shaft for producing a current of air for ventilating the mine. The power of a *furnace* where the shafts are 600 yards deep and over, is probably greater than that of a *fan* as ordinarily constructed. As much as 400,000 cubic feet of air per minute have been passed up a single shaft by furnace ventilation. It has its disadvantages, however, viz. the chief being, the liability of sparks from it to ignite an explosive mixture in the *upcast*, and thereby cause an *explosion* in the mine attended with terrible consequences. The excessive heat in the shaft, rendering it in many cases unfit for *winding* in, or for any other than ventilating purposes. The liability of the fires to get low through the negligence of the *furnaceman*. Of the heat of the furnace to set fire to the coal, &c., in the locality; of the shaft-fittings to take fire; the *tubbing*, &c., to become dangerously weak from the effects of heat, wet, &c.

FURNACEMAN. One whose sole occupation is to keep the *furnace* going.

FURTHERANCE (N.). An additional sum of money paid per *score* to *hewers*, *putters*, &c., as an allowance in respect of inferior coal, a bad *roof*, a *fault*, &c.

FUSE or FUZE. A small train of gunpowder enclosed in a hollow cord of hemp, &c., for firing off *shots*.

## G.

GAD. An iron wedge used for breaking down coals, &c.

GAGING (S. S.). A small embankment or heap of slack or rubbish, made at the entrance to a *heading*, &c., as a means of fencing it off.

GAGS. Chips of wood in a *sinking pit* bottom, or *sump*.

GAILLETINS (Belg.). Round coal.

GAIN (M.). A transverse channel or *cutting* made in the sides of a roadway underground for the insertion of a *dam* or close permanent *stopping*, the object being to prevent any *gas* escaping or any air entering, and to retain the *dam* in a firm position.

GALE (F. D.). A specified tract of mineral property granted by the Crown to a colliery proprietor or company for working the mines.

GALEE (F. D.). The owner of a *Gale*.

GALLOWS (N.). A *crown tree* with a prop placed underneath each end of it. See Fig. 59.

GANG. 1. (M.) To go; to move along.

2. A train or *set* of pit *tubs* or *trams*.

GANGER (M.). One who is employed at conveying minerals along the gangways in or about a mine, which employment is known as *ganging*.

GANG-RIDER. A lad who rides with or upon the *trams* upon underground *engine planes*, to give signals when necessary, and to work any *clips*, &c. See *Haulage Clip*.

GANGWAY (Pa.). The main haulage road or level, which is driven on the *strike* of the mine.

GANNEN (N.). A *board* down which coals are conveyed in *tubs* running upon rails.

GANNISTER. A very hard and compact, extremely siliceous *fire-clay*, being the *floor* of some of the lower coal seams of the Midland coalfield. It is often crowded with the fossil *Stigmaria*, and is largely made use of for lining the interiors of steel furnaces, converters, &c.

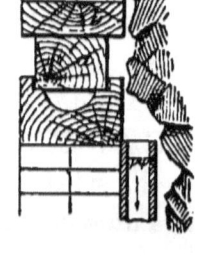

Fig. 73.

*a*. small blocks of wood placed at intervals round the curb to support the upper ring *b*.

GARLAND. 1. A wooden or cast-iron *curb* set in the *walling* of a *pitshaft* to catch and conduct away into a pipe or *lodge*, any water which runs down the shaft sides. See cross section of a *garland* or *water curb*, Fig. 73.

2. A wooden frame, rectangular in form, and strengthened with iron corner-plates, for

keeping the coals together upon the top of a *tram*, &c., when heavy loading is practised in a mine. Sometimes two, and even three are used upon one load. See end view, Fig. 74.

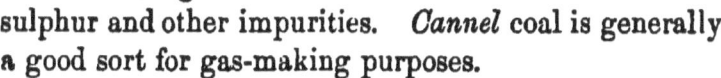
Fig. 74.

GAS. See *Fire-damp*. Generally any mixture of this gas and air in an explosive condition is called *gas*.

GAS COAL. That which yields a large quantity of illuminating gas on distillation, together with freedom from sulphur and other impurities. *Cannel* coal is generally a good sort for gas-making purposes.

GAS DRAIN. A heading driven in a mine for the special purpose of carrying off or draining away *fire-damp* from a *goaf* or other working. Sometimes a bore-hole put down from an upper to a lower seam of coal with a similar object, or a bore-hole put into the floor to liberate gas, which is known in some places to exist in coal under the enormous pressure of over 300 lbs. to the square inch.

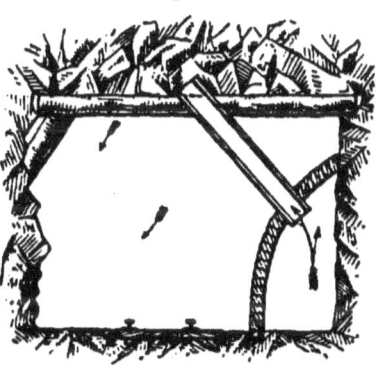
Fig. 75.

GAS-MAN (U. S. A.). See *Fireman, Fire-bosses*.

GAS-PIPE (M.). A short wooden pipe about 4" × 4" inside, having its upper end open to the roof in the cavity to which it is applied, and the lower end opening into the *bratticing* (see Fig. 75), so that any gas given off in the roof is, by the air drawn up the pipe, diffused and carried away as formed, and no fall of roof at that point can suddenly force out gas previously accumulated, upon *naked lights*.

GATE (from the Saxon verb *Gangum*, to go). An underground road connecting a *stall* with a *main road* or inclined plane, worked either by horses and ponies or by self-acting incline ropes or chains.

GATE-END. The *inbye* end of a *gate*.

GATE-END PLATE (M.). A large iron plate or sheet about 4′ 6″ square and ½″ thick, upon which *trams* are turned round upon coming out of the *stall face* to be taken along the *gate*. Smaller plates are sometimes used, one laid between the tram-rails and one on either side of it.

GATE-ROAD (M.). See *Gate*.

GATE-WAY (M.). See *Gate*.

GATHER (D.). To drive a *heading* through disturbed or *faulty* ground in such a way as to meet with the seam of coal, &c., sought, at a convenient level or point on the opposite side. See *Eat-out*.

GAUGE-DOOR. A wooden door fixed in a mine in an airway for regulating the supply of ventilation necessary for a certain district, or number of men, &c. Its opening is adjusted by various means, and is solely controlled by the *underviewer* or *manager*.

GAUTON (S.). A narrow channel or *ricket*, cut in the floor of an underground roadway.

GAUZE LAMP (S.). A (so-called) safety-lamp, formerly commonly used in the Scotch coal-pits. It is a kind of *Davy* lamp, with a gauze top about 3 inches in diameter, and has no brass frame to strengthen it, and no glass.

GAVELLER (F. D.). The Crown agent, or *gale giver*, who has power to grant *gales* to *free miners*.

GAWL (L.). An unevenness in a coal *wall*.

GAYETTE (Belg.). Large picked coals.

GAYLETTERIE (Belg.). Second quality coals.

GEAR (N.). A collier's tools, consisting of *picks*, drills, wedges, hammer, shovel, &c.

GEARS. 1. (N.) See *Double Timber*.

2. (N.) Staging and rails erected at quays over *coal drops*.

GEODES (Lei.). Large nodules of *ironstone*, hollow in the centre.

GEORDIE. A safety-lamp invented by "the father of the railway system" (George Stephenson) in 1815. He, although quite independently of Sir H. Davy (who also invented a safety-lamp, the *Davy*), is said to have been the first to produce a lamp which would indicate an explosive mixture of gas and air in a mine without causing an *explosion*. The *Geordie* lamp is extinguished by the presence of *firedamp*. The flame of this lamp is surrounded by a glass cylinder fitted with a perforated metal cap, a wire gauze cylinder forming the outer or essential part of the arrangement. The gas enters the lamp through a number of small holes in the base of the lamp-top, takes fire at the flame, and the *after-damp* (the products of combustion) puts out the light. It, however, gives a miserable light, and is unsafe when exposed to a high velocity in an air-current charged with much gas.

GERMAN. A straw filled with gunpowder to act as a *fuze* in blasting operations.

GET. 1. To work away or excavate by mining either under or above ground.

2. The produce or *output*, in tons, of a colliery or mine during a certain period, e. g. 125,000 tons in six months.

GETTING. Cutting, falling, and loading up of the coals, &c., in a mine.

GETTING ROCK (S. S.). *Clay ironstone* in the *roof* of a coal-seam, which is worked in conjunction with the coal.

GHOST (S. S.). A *blue cap* on a candle or lamp.

GIB. A short prop of timber by which the coal is supported whilst being *holed*, or undermined. See *Sprag*.

GIN or HORSE GIN. A *drum* and framework carrying small pulleys, &c., by which the minerals and dirt are raised from a shallow pit, not exceeding say 35 yards, or from a dip incline from surface, or one in the workings. A *gin* is also used for raising the materials, &c., in building tall chimneys, &c.

GIN-BEAM (S. S.). A timber cross-bar carrying the pulley-wheels over the top of a *gin-pit*.

GINGING (D.). The *walling* or lining of a *pit-shaft*.

GINNEY. See *Jinney*.

GIN-PIT. A shallow mine or a *pit-shaft*, say from 10 to 35 yards deep, worked by a *gin*. The coal is hoisted in small wooden tubs or boxes without wheels, carrying about 3 cwt. each, and swinging loose in the pit-shaft, one up and one down.

GIN-RACE or GIN-RING. A wide excavation near the top of an underground inclined plane to the *dip* in which a *gin* is fixed. When on the surface it means the circular space occupied by the *gin, &c.*

GIRDLES (N.). Thin beds of sandstone, &c., exposed in a *sinking-pit* or in a *bore-hole*.

GLANCE COAL. Another term for *Anthracite*, which see.

GLASS. A collier's word for a *dial*.

GOAF, or GOAVE. That part of a mine from which the coal, &c., has been worked away and the space more or less filled up. See *Double Stall*, Fig. 58; also *Head* (8), Fig. 80.

GOB. 1. Another word for *Goaf*.

2. To leave behind in the mine coal and other minerals which are not marketable.

3. To stow or pack full of rubbish any useless underground roadway.

GOBBIN or GOBBING (Lei.). See *Goaf*.

GOB-FIRE. Spontaneous combustion underground. It would seem in a great measure to be due to the action of iron pyrites becoming oxidized by the co-operation of moisture. During the decomposition the coal becomes split up, and exposes a larger surface to the air; the ferrous salt is then oxidized into the ferric salt, which gives up its oxygen to the coal. In order to prevent *gob-fires* it would appear necessary to exclude all currents of air, unless passed through the place from the commencement in a strong current, so as to act as a cooling agent.

GOB ROAD. A gallery or *way* in the mine carried through a *goaf*. Many seams of coal, &c., are worked by what is known as the *gob-road system*—that is to say, all the main and branch roadways are made and maintained through the exhausted portions of the mine, the regular workings in which are opened out and carried forward from the sides of the *shaft-pillar*. Mines worked upon the *longwall* system are generally worked *gob-road*, particularly in the Midland counties of England, where the mines are very flat.

GOB-WALL (S. W.). A rough kind of wall constructed of the stone from the *roof*, &c., built up and carried on along either side of a *gob road* in order to keep up the roof and maintain a good roadway through the pit.

GOING. Being worked forward or advanced in any direction, e.g. *headings* in course of being worked or cut are said to be *going*.

GOING BOARD (N.). A board down which coals are trammed, or one along which the *stuff* from several working places is conveyed into the main *wagon-way*.

GOOSE (F. D.). A water-barrel or tub.

GOSKINS.

GOT-ON-KNOBS (S. S.). A system formerly practised of working the Thick coal, being a kind of *board and pillar* plan, the main roadways being first driven up to the boundary.

GOTTEN (M.). Worked out or exhausted *mine* (1 and 2).

GOUTWATER (F. D.). Mine water containing sulphuretted hydrogen.

GOWL (D.). *Roof* and *sides* are said to *gowl* or *gowl-out* when they break down and cause trouble.

GRABS (Pa.). A tool for extricating broken boring tools out of a *borehole* (1), consisting of two iron side-rods fitted at the lower ends with half arrow-headed points facing inwards.

GRAFTING SPADE. A long narrow-plated spade for digging clay.

GRAITH (S.). Tools used by a *collier* (1).

GRAPIN (F.). A tool used in the *Kind-Chandron system* of *sinking shafts*. It is in form like a gigantic pair of scissors, the points of which cut away and trim up the edges of the shaft in preparing a seat or bed for the *moss-box* to rest upon.

GRAPPEL. A cutting tool for obtaining a solid specimen of the rock bored into. See *Carrot*.

GRASS. The surface. The pit bank (1). The expression "gone to grass" means gone up the pit or gone to *bank* (1).

GRATHE (N.). To replace, repair, dress, or put in order.

GRATHELY (N.). Tidy, orderly.

GRATHER (N.). See *Changer*.

GRAVEL WALL (W.). The junction of a coal-seam with overlapping or uncouformable Permian, &c., rocks.

GREEN ROOF. A miner's term for a *roof* which has not broken down or *weighted* at all.

GREYS (Som.). Hard siliceous sandstone.

GRIDAW (S. W.). *Pulley Frames* or *Head Gear*, which see.

GRIMES (S. W.). See *Bell-mould*.

GRIST (S. W.). A black coaly stratum indicating a probable vein of coal not far off.

GRISOU (F.). See *Fire-damp*.

GRIZZLE. Inferior coal with an admixture of specks and patches of iron pyrites, and often sooty.

GROS MORCEAUX (Belg.). Coal in very large lumps.

GROUND. Strata or *measures*. When strata do not contain coal or other mines of sufficient thickness or value to make them *workable* at a profit, they are said to be *barren* or *unproductive ground*. The terms *hard ground, soft ground, faulty ground, broken ground*, &c., are very commonly made use of.

GROUND BAILIFF (M.). Old term for *Manager*. His duties were to look after the getting and sending to *bank* (1) of the coal, keep the ventilation right, &c.; but had generally nothing to do with the machinery or mechanical department of the colliery.

GROUND BLOCKS. Pulley blocks to which the *ground spears* are hung.

GROUND CRAB. A species of capstan used for lowering the *sinking set* of pumps as the shafts get deeper.

GROUND RENT. Rent paid for surface occupied by the plant, &c., of a colliery; generally double the usual agricultural or surface-rent.

GROUND ROPES. Hemp ropes for passing through the *ground blocks* to the *ground crabs*.

GROUND SPEARS. Wooden pump-rods (one on each side of the *set* or *pump trees*), to which the pumps in a *sinking-pit* are suspended.

GROWL (M.). Coal *pillars*, &c., are said to *growl* when they are undergoing a crushing weight.

GUELL (I.). Coal.

GUG (Som.). A self-acting inclined plane underground; sometimes a *dip* incline.

GUIDES. 1. See *Cage Guides*.

2. A *boring-rod* having an enlargement or wings fitted to it to suit the size of the *borehole* (1) for steadying the rods when a considerable depth has been attained.

GUIDING BED. A thin band or seam of coal, &c., in a *nip* leading to the regular seam on either side of it. See Fig. 70 (2).

GULCHING (N. S.). The moving and crackling noise made by a *weight* coming on underground.

GUM (S.). Free-burning small slack or *duff*.

GUNBOAT (Pa.). A *car* or wagon holding from 5 to 8 tons of coal, used upon inclined planes or *slopes*. They are filled by emptying the *trams* into them at the foot of the slope, and empty themselves on reaching the surface, when the coal runs down on to *screens* for separation and cleaning.

GUSS (B.). A short piece of rope by which a boy draws a *tram* or *sled* in a pit.

GUTTER. 1. (F. D.) An *air-way* through a *goaf*.

2. Candles or dips, when subjected to the warm air of a mine, waste away very rapidly, and are said to *gutter* or *sweal*.

GUTTERING (Pa.). A channel or pipe cut along the side of a *pit shaft* to conduct the water not *tubbed back* into a *lodge* or *sump*.

GUTTER-UP (M.). See *Cut-up*.

GUYS. Strong wire ropes or cables attached near the top of *headstocks*, and anchored at the ground to keep them steady.

GWYTHYEN (S. W.). A *vein* or *seam*.

# H.

H-PIECE. A strong pipe cast in the form of a letter H containing the bottom *clack* of a forcing *sett* (1) of pumps. One side communicates with the *plunger*, the other with the suction and delivery, and has a *clack door* on it. See Fig. 76.

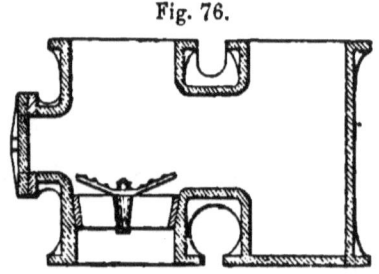

Fig. 76.

HACK (N.). A *pick* or tool with which colliers cut or hew the coal, and use in *sinking* and *stone drifting*. It weighs about 7 lbs.

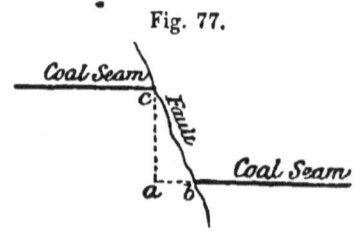

Fig. 77.

HADE. The dip, inclination, or underlie of a *fault*, measured by the angle between a vertical plane and the plane of the fault. In Fig. 77 the dotted line *a b* represents the *hade* as distinguished from the *throw* or amount of displacement which is the length of the line *a c*.

HAGGED (S.). Hewn or cut.

HALF-COURSE. Half on the level and half on the dip.

HALF-END. See *Horn Coal*.

HALF-END AND BOARD (Y.). See *Horn Coal*.

HALF-MARROW (N.). A *butty* or partner.

HALF-MOON. A scaffold nearly filling up one half the sectional area of a *pit-shaft*, or in plan the form of a half-moon, upon which repairs are done.

HAND DOG. A kind of spanner or wrench for screwing up and disconnecting the joints of *boring rods* at the surface.

HANDFUL (B. and Som.). A length of four inches.

HAND or HANDLE. To work a winding, pumping, hauling, or other engine.

HANDLING (M.). Reloading coals underground from one *tub* to another.

HANG (B.). The lie or *hade* of a *fault*.

HANGER ON. The man who runs the full *trams* upon the *cages* and gives the signals to *bank* (1).

HANGING ON. The pit bottom, level, or *inset*, at which the *cages* are loaded.

HANGING SPEAR-RODS. Wooden pump-rods adjustable by screws, &c., by which a *sinking sett* of pumps is suspended in a *shaft*.

HARD-HEADING. A *heading, tunnel,* or *drift*, driven in *stone* or *measures*.

HARDS (M.). Coals of a hard and close-grained character.

HARP (S.). To fill a *hutch* with coal at the *face*.

HATCH (B.). See *Door*.

HATCHING (B.). An underground way or self-acting inclined plane, in a thin seam of coal, carried up from 60 to 80 yards to the *rise*.

HAT ROLLERS. Cast iron or steel rollers, shaped like a hat, revolving upon a vertical pin, for guiding incline hauling ropes round curves. See Fig 78.

FIG. 78.

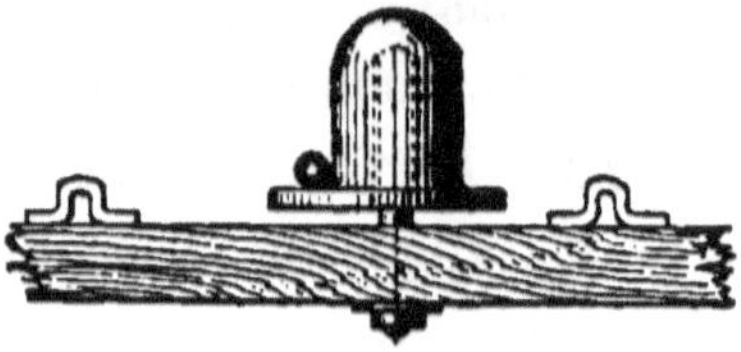

HAULAGE or HAULING. The drawing or conveying of the produce of the mine from the *working places* to the bottom of the *winding* pit. This work may be performed in the following ways :—By pushing the *trams* by hand, as is done in very small pits ; by horses or ponies drawing several trams at a time ; by self-acting inclined planes driven of course to the *rise*; by stationary engines worked by steam, compressed air, or hydraulic power working wire ropes, or chains, and by locomotives working with compressed air. In most mines some kind of mechanical *haulage* is to be found, but horses are invariably used as well, to convey the trams from the *stalls*, &c., on to the *main roads*. Hauling coals a distance of about three miles is occasionally performed. Horses to the number of 80 are sometimes to be found assisting in hauling in one colliery, and over 2000 tons of mineral are sometimes conveyed to the pit bottom in one day.

USED IN COAL MINING, ETC. 129

HAULAGE CLIP. Levers, jaws, wedges, &c., by which *trams*, singly or in trains, are connected to the hauling ropes. There are several ingenious and simple arrangements in use, some of which are given in Fig. 79.

Fig. 79.

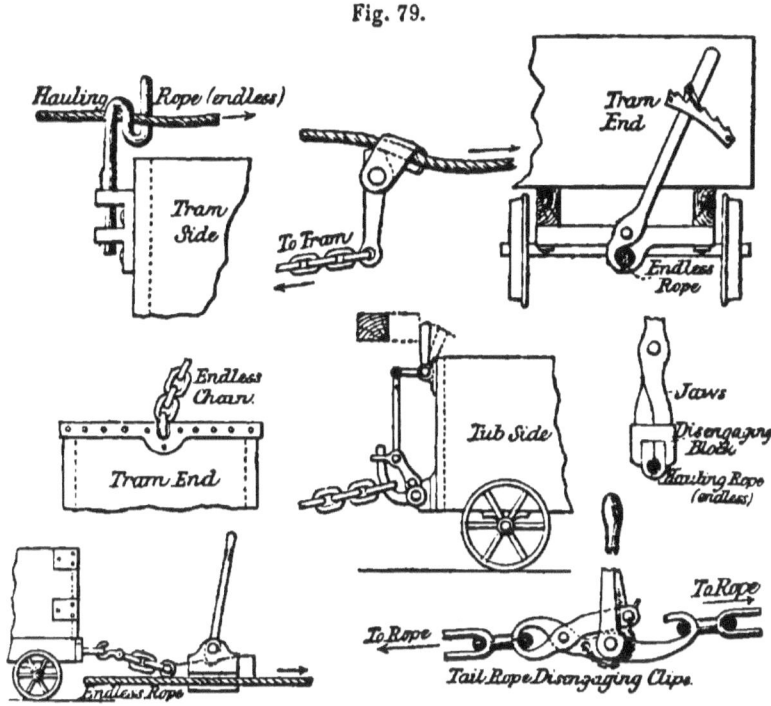

HAULIER. A boy or man who goes with a pony or horse in the pit, or who attends the trains upon engine planes, &c.

HAUNT (Som.). See *Landsale*.

HAZLE (N.). A tough mixture of sandstone and shale.

HEAD. 1. Any road, level, or other subterraneous passage driven or formed in the solid coal, &c., for the

K

purpose of proving and working the mine. A common size for an ordinary *head* is 6 feet by 6 feet, though the sectional area depends entirely upon circumstances, some being as much as 70 to 80 square feet, the smallest about 8 or 10 square feet.

2. That part of a *face* nearest to the *roof*.

3. (Som.)  Any length of working *faces*.

4. (S.S.)  A shift or day's work by the *stint* in *heading* (2) out, or driving of *deadwork*.

5. The top end of the *boring rods* above the surface.

6. Pressure of water in pounds per square inch, or, of so many feet.

7. To cut or otherwise form a narrow passage or *head* (1).

8. A *lift* (3).  See Fig 80, showing a seam being worked in three *heads*.

Fig. 80.

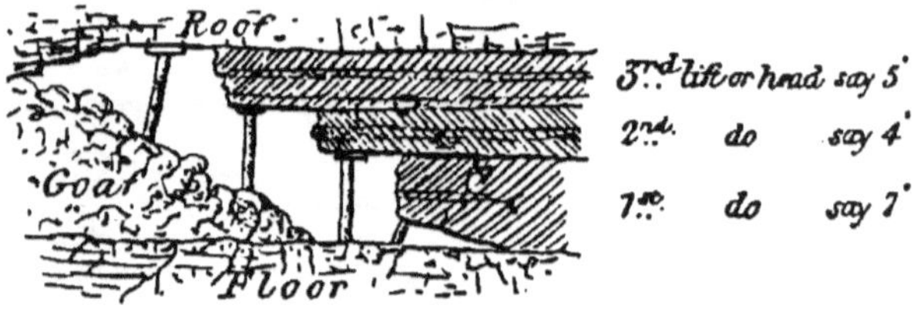

3rd lift or head say 5'
2nd.  do    say 4'
1st.  do    say 7'

9. See *Motive Column*.

HEAD-COAL (S.). The upper portion of a thick seam of coal which is worked in two or more *lifts* (3).

HEADER (M.). A *collier* or coal cutter who *drives* a *head* (1); he is paid by the yard and also receives so much per ton upon the large coals sent out.

HEAD-GEAR. The pulley-frame erected over a *winding shaft* constructed of iron or timber or both, and sometimes reaching to 72 feet in height. For boring work it is generally from 30 to 40 feet high, though as much as 80 feet are occasionally employed, Norway fir being the kind of timber used.

HEADING. 1. See *Head* (1).

2. The operation of driving a *head* (1).

3. (Pa.) A level driven parallel to a *gangway* and usually the *return* airway of the mine.

4. (S.) The top portion above the *tub* sides of the load carried.

HEAD-ROOM. Height as between the *floor* and the *roof* anything above 6 feet is considered good *head-room* in a pit.

HEAD-SIDE (N. S.). The *rise* side of a *heading* (1) driven on the *strike*.

HEADSMAN (N.). A *putter* or *haulier*, which see.

HEADSTOCKS. See *Headgear*.

HEAD-TREE (N.). A portion of a *crown-tree* about 12 inches in length.

HEADWAYS (N.). The direction of the *cleat* or a *place* (1) driven parallel with the *cleat*, that is, *end-on*.

HEADWAYS COURSE (N.). When a set of *headings* or *walls* extend from side to side of a set of *boards* they are said to be driven *headways course*.

HEAP (S.). To load up a *tub* above the top of the sides.

HEAP-KEEPER (N.). The head *banksman* who looks after the sorting and cleaning of the coals, and keeps order about the pit top, &c.

K 2

HEAP-STEAD. The entire surface works about a colliery shaft; includes the headgear, loading and screening arrangements, winding and pumping engines, &c., with their respective houses. The workshops, stores, &c., being sometimes built into the same block surrounding the pit top. Fig. 81 is a plan of a *Heapstead* of a large colliery.

HEAT. The elevated temperature produced by spontaneous combustion.

HEATH or YERTH (S. S.). Earth.

HEAVE. 1. See *Creep*.
2. A *fault* of dislocation.

HEAVY. The hollow sound produced when knocking on a *roof*, &c., which is giving way. An unsound or dangerous roof is said to *knock heavy*.

HEAVY FIRE (N.). An extensive and severe *explosion*.

HEIVER. A coal cutter or *hewer*.

HELVE or HELVER. The handle of a *pick* or *maundrill*.

HESS (S. S.). Clinker from furnaces of boilers.

HEUGHS or HEUCHS (S.). Ancient term for coal seams or coal workings.

HEWER. A collier who *cuts* coal.

HIGH PILLAR. See *Shaft Pillar*.

HILL (N. M.). An underground inclined plane.

HINGING (Y.). See *Cap, re Ropes*.

HIT. To find, prove, or cut into a coal seam, fault, &c.

USED IN COAL MINING, ETC. 133

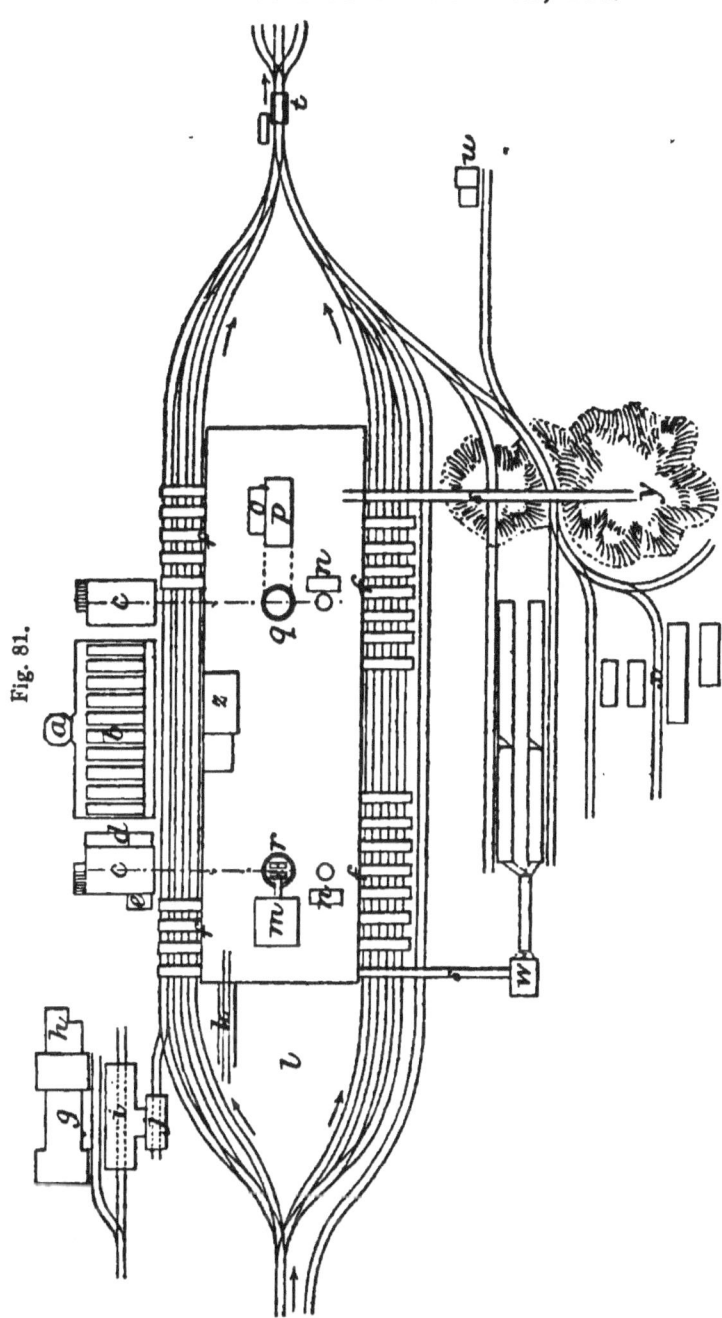

Fig. 81.

*a*, Chimney or stack. *b*, Boilers. *cc*, Winding engines. *d*, Donkey. *e*, Steam crab. *ffff*, Screens. *g*, Workshops, &c. *h*, Stores. *i*, Wagon shop. *j*, Loco. shed. *k*, Timber incline to pit bank. *l*, Timber for underground use. *m*, Pumping engine. *nn*, Weighing machine. *o*, Fan engine. *p*, Fan. *q*, Upcast shaft. *r*, Downcast shaft. *ss*, Dust road to coke ovens. *t*, Truck weighing machine. *u*, Powder magazine. *vv*, Coke ovens. *w*, Washing apparatus. *x*, Brickworks and Gasworks. *y*, Spoil bank. *z*, Air-compressors.

HITCH. 1. (S.) A *fault* of dislocation of less *throw* than the thickness of the seam in which it occurs.

2. (S. W.) To attach *trams* to hauling ropes by short chains, &c.

HITCH AND STEP (S. W.). A system of regulating the distance between the *faces* of *stalls* in *long-wall* work. See Fig. 82.

Fig. 82.

HITCHER. The man who runs *trams* into or out of the *cages*, gives the signals at *bank* (1), and attends at the shaft when men are *riding* in it.

HOD (F. D.). A *cart* or *sled* for conveying coals in the *stalls* of *thin seams*.

HOG-BACK. Sharply rising of the *floor* of a coal seam.

HOGGER. 1. (N.) Stockings without feet, chiefly worn by *hauliers*.

2. The uppermost pipe of a pumping *sett*, through the side of which water is discharged through a wide leather pipe.

HOIST. An elevator or lift, either single or double acting, worked by steam or hydraulic power, for raising the tubs of coal on to the screening stage from the *bank* (1) level.

HOLD OUT! (D.) This was shouted by the *banksman* down a *pit-shaft* to the *bottomer* when a *bant* of men were about to descend the *shaft*, to let him know that he was not to send up a load of coals against the *bant*, but merely the *empty rope* or chain, in order to avoid accident by collision known as a *wedding*, which see.

HOLE. 1. To undercut a seam of coal, &c., by chipping away the coal, &c., with a *pick*, or by the employment of a machine worked by compressed air to do the same work.

2. A *borehole*, which see.

3. To make a communication from one part of a mine to another.

HOLES (N.). The different *flats* (1) or stages from which the *cages* are loaded at the pit bottom.

HOLES OF SAWYER (S.S.). Blocks of the Thick or Ten-yard coal-seam formed by *holing*, and then cutting the sides upwards by forming vertical grooves between the mass to be brought down and the sides of the *pillars* to be left unwrought to support the *roof*. [The term *sawyer* refers to a particular band or layer forming portion of the Thick coal.]

HOLING. 1. The wedge-shaped portion of a *seam* or *floor* removed from beneath the coal before it is broken down. Sometimes the *holing* is made in the top of the seam, sometimes in or about the middle. It is only in hard or moderately hard coals that *holing* to any considerable depth or distance under is necessary; but in order to produce coals in the best possible shape or size deep *holing* is indispensable. A hard seam should be *holed* to a depth of not much less than the thickness of the seam, e. g. a six feet seam holed five to six feet under. See *Bannocking*.

2. A short passage connecting two roads.

HOLLOW REAMER (Pa.). A tool for straightening a crooked *borehole* (1).

HOLLOWS. Old abandoned *workings*.

HOME (N.). In the direction of the shafts. When a certain quantity of air has circulated through a sufficient length of workings it is sent *home* or direct to the *upcast*.

HOO CANNEL. Impure earthy *cannel* coal.

HOOKER ON. See *Hanger on*.

HOPES (N.). Valleys formed by denudation in the *coal measures* of the County of Durham.

HOPPITT. See *Bowk*.

HORN COAL. Coal worked partly *end-on* and partly *face-on*. This is the proper way to work a hard seam to the best advantage.

HORN-SOCKET (Pa.). See *Bellscrew*.

HORSE. See *D-Link*.

HORSE-BEANS (Ch.). A stratum of a granular structure immediately overlying the rock salt beds, in which the *rock-head* brine runs.

HORSE-FETTLER (S. S.). A man who looks after the underground horses and ponies.

HORSE GIN. See *Gin*.

HORSE-HEIGHT (M.). Distance between the *floor* and the *roof*, for a horse to travel without knocking his head, &c.

HORSE-LOAD (L.). A measure of weight used in some parts of East Lancashire. 1 horse load = 4 cwt. or 5 horse loads to a ton.

HORSE-ROAD. An underground *way* worked by *horsing*.

HORSES or HORSEBACKS. Natural channels cut, or washed away by water, in a coal seam, and filled up with shale and sandstone. Sometimes a bank or ridge of foreign matter in a coal seam.

HORSE-TREE. A strong timber beam to carry pumps, &c.

HORSING. Drawing trams underground by horses and ponies.

HOUSE, HOUSE-FIRE, HOUSEHOLD COAL. Has a hard fracture and in burning leaves little ash, and that of a reddish-brown colour.

HOWDIE HORSE (N.). A pit horse kept on the surface for use in cases of emergency.

HOW WAY! (N.) Lower the *cage* down.

HUDDOCK (N.). The cabin of a *keel*, which see.

HUDGE (Som.). See *Bowk*. Also a small *box* or *tram* without wheels running on timber slides, drawn by a boy in thin and steep seams.

HUGGER (N.). A *Back* or *Cleat*.

HUNCH (D.).

HUND (Pr.), meaning *dog*. A rectangular iron tram or wagon on four small wheels with a projecting pin beneath it to run between the rails (wooden), and thus guide the movement. Used as long ago as 1550.

HUNDRED. Hundredweight (cwt.).

HUNKER (In.). Yellowish clay containing concretionary nodules.

HUNTING COAL (Y.). *Ribs* and *posts* of coal left for second working.

HURDLE SCREEN (S.). A temporary screen or curtain for clearing *gas* out of a pit.

HURLEY (S.). A *Hutch*.

HURRIER. See *Haulier*. Generally small boys.

HURRY. To haul, pull, or push *trams* of coal, &c., in a mine.

HUTCH (S.). See *Box*.

HUTCH RUNNER (S.). Boy who draws *hutches*.

HYDRAULIC PUMPING ENGINE. An apparatus using water as its motive power, for draining such portions of the underground *workings* as are below the level, or to the *dip* of the *shafts*; for pumping water up a *shaft* to the surface pumping engine, or to a steam engine placed part way down the *shaft*. The principle of its action is that of employing water at a given *head* (6) to raise a larger quantity against less *head*.

# I.

IN. When a *stall* or other *working place* in a mine is blocked up with fallen *roof*, &c., it is said to be *in*, or to have *come in*.

INBYE. Going into the interior of a mine, away from the shafts or other openings. Fresh air and empty tubs go *inbye*.

INCLINE. 1. Short for Inclined Plane. Any underground roadway which is driven at an angle to the horizon. If to the *rise* it is worked by a self-acting arrangement, if to the *deep* by a steam or other engine.

2. To dip sufficiently to form a self-acting *incline* (1).

INCLINE ROPE HAULAGE. A system of *haulage* in which a single rope is used, or where the inclination of the *plane* is such as to allow of the empty *tubs* drawing the rope in after them.

INCLINE DRAW-ENGINE. A stationary surface inclined-plane engine.

INDICATOR. 1. A mechanical contrivance attached to winding, hauling, or other machinery which shows the position of the *cages* in the *shaft* or the *trams* upon an *incline* during its journey or run.

2. An apparatus for showing the presence of *firedamp* in mines. The temperature of *goaves*. The speed of a *ventilator*, &c. And also for calculating the power of an engine.

IN-DOOR CATCHES. Strong beams in Cornish pumping engine-houses, to catch the beam in case of a smash, and prevent damage to the engine itself.

IN-DOOR STROKE. That stroke of a Cornish pumping engine which lifts the water in the bottom or drawing *lift*.

IN FORK. When pumps are working with the water having receded below some of the holes of the *windbore*, they are said to be *in fork*.

INGATE (N.). See *Inset*.

IN-GOING. That which is going *inbye*.

IN-OVER. See *Inbye*.

INSET. The entrance to a mine at the bottom or part way down a shaft where the *cages* are loaded. See Fig. 69.

INSPECTOR. 1. (N.) A man appointed to overlook the *banking* and *screening* department.

2. Her Majesty's Inspector of Mines, of whom there are several.

INTAKE. 1. The fresh air *airway* or road going *inbye*, commencing at the bottom of the *downcast*.

2. The fresh air descending into a colliery.

INTERBEDDED. When patches or layers of strata or of *trap* (having no true relation to the coal measures) lie between two beds, the rocks are said to be *interbedded*, e. g. the sheet of intrusive dolerite in the Leicestershire coal-field.

IRON MAN. A collier's term for a coal-cutting machine.

IRONSTONE. A term usually applied to argillaceous or *clay ironstone*, containing from 20 per cent. to 40 per cent. of iron. It is very commonly met with in the *coal measures*, and takes the form of thin beds or layers and of nodules or balls of various sizes and shapes—is interstratified with the shales and clays throughout the entire series of the measures. Sp. gr. about 3. A cubic foot weighs from 170 to 190 lbs. The ironstones or ores of the Lias and Oolite series of rocks are found in beds as much as from 10 or 20 feet thick, these ironstones are of less specific gravity than the *clay* or *blackband* varieties. Great Britain produces annually something like 15,000,000 tons of ironstones of various kinds.

IXOLITE. A mineral found in certain bituminous coals.

## J.

JACK 1. (N.) A lantern-shaped case made of tin in which *safety lamps* are carried in strong currents of ventilation.

2. (S.) One who works underground at *odd work*.

JACKANAPES. The small guide pulleys of a *whim*.

JACK ENGINE (N.). The engine for raising men, débris, &c., in a *sinking pit*.

JACK HOLES (N. S.). See *Cut through*.

JACK LAMP. A *Davy* lamp with the addition of a glass cylinder outside the gauze.

JACK PIT (N.). A shallow *pit-shaft* in a mine communicating with an *overcast*, or at a *fault*.

JACK-ROLL. A windlass worked by hand.

JACKS. 1. (N.) Large fissures or cracks in the *roof*.

2. (Lei.) Wood wedges 6" × 4" tapered at one broad edge, so that when driven up they cannot start again.

JACKY PIT. See *Jack Pit*.

JAD (Som.). A long and deep *holing, cutting*, or *jud*, made for the purpose of detaching large blocks of stone from their natural beds at the Bath-stone (Oolitic) quarries, or rather underground workings, at Box.

JADDING. The operation of forming a *jad*.

JADDING PICK. The tool employed to cut a *jad*. They are made in sets of about three or four, with *helves* ranging from three to six feet in length, to enable the *jads* to be cut to a great depth.

JAILER (Som.). A small tub or box in which water is carried in a mine.

JAM OUT (S. S.). To cut or knock away the *spurns* in *holing*.

JARS (Pa.). A sliding joint in *boring rods* for deep holes, consisting of two long loops of iron or steel, sliding one within the other.

JAY (D.). Roof coal.

JENKIN (N.). An opening cut into or a slice taken off a *pillar* from six to eight feet in width, in the *board and pillar* system of working coal.

JET. A compact, black, lustrous, resinous variety of *lignite*, susceptible of a high polish. It occurs chiefly in the Upper Lias clays of Yorkshire, &c., in lenticular patches or beds, nodules, and irregularly shaped masses. Is believed to be formed of the fossilized stems of coniferous trees. The Romans used it. Some 1500 hands are employed in the *jet* trade (mining, cutting, polishing, &c.), and the value in 1872 is stated to have been 88,000*l*. Jet is mined by driving levels and systematically exploring the strata by a kind of stoping or overhead excavating. English jet is worth from 300*l*. to 1300*l*. per ton.

JIDDY (L., N. S.). See *Runner* (1).

JIG. A self-acting *incline* worked by a *drum* (2) or by wheels, with hemp or steel wire ropes. Fig. 83 shows a useful and inexpensive arrangement for light loads and short runs.

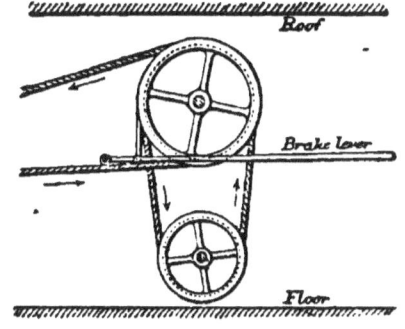

Fig. 83.

JIGBROW (L.). See *Jig*.

JIGGER. 1. (S.) A kind of coupling hook for connecting *trams*, upon an incline.
2. (Lei.) See *Onsetter*.

JIG RUNNER (Y.). The man who works a *jig*.

JINNEY. See *Jig*.

JINNEY TENTER. See *Jig Runner*.

JITTY (Lei.). A short *slit* along which *empties*, horses, or workmen travel.

JOCKEY (M.). A self-acting apparatus carried on the front *tub* of a *set*, for releasing it from the hauling rope at a certain point. See Fig. 84.

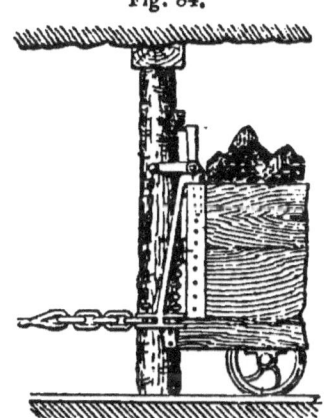

Fig. 84.

JOEY (M.). A man specially appointed to set *timber* in a *stall* during the *turn* (1). He is a *butty*, and is not paid for doing this work, but takes his turn at it with the other *butties* in his *stall*.

JOINTS. Natural divisions, cracks, or partings in strata.

JOURNAL. A carefully kept diary, schedule, or statistical account of the various operations connected with the putting down of a *borehole* (1) in search of coal, &c. The following arrangement of the page for such a book may be taken as a guide in preparing the *journal*; it is taken from the work on 'Mine Engineering' by G. G. André.

| Date. | Description of Strata. | No. of Specimen in Case. | Thickness. | Depth from Surface. | Angle of Dip. | Diam. of Hole. | Description of Tool employed. | Time actually occupied in passing through. | Quantity of Water met with. | Organic Remains. | Remarks. |
|---|---|---|---|---|---|---|---|---|---|---|---|

JOURNEY (S. W.). A *train* or *set* of *trams* all coupled together running upon an engine plane: as many as forty sometimes.

JOWL or JOWELL (N.). See *Chap*.

JUD. 1. (N.) A block of coal about four yards square *kirved* and *nicked* ready for breaking down.

2. (Som.). See *Jad*.

JUDGE. A staff used for gauging the depth of the *holing*. Formerly a boy who proved the *holing*. Fig. 85.

Fig. 85.

JUDGE-RAPPER. The upper end of the vertical arm of a *judge*. See Fig. 85.

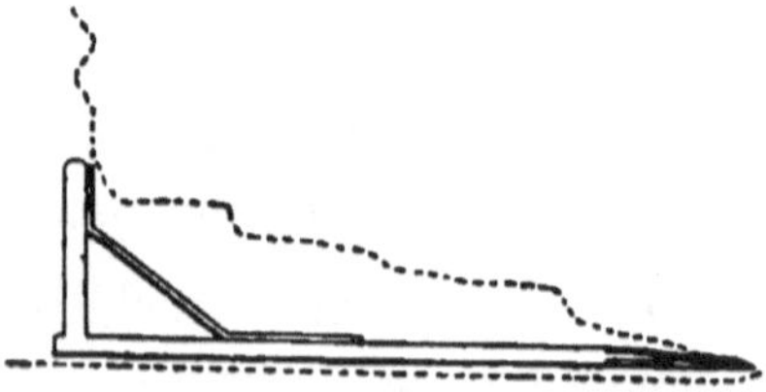

JUMP (Jump-up, Jump-down). 1. An *up-throw* or a *down-throw, fault*.

2. To raise *boring-rods* in a *bore-hole* (1) and allow them to fall of their own weight.

JUMPER. A hand drill used in blasting, having at each end a chisel edge and a swell or bead in the middle to give it more weight.

JUNKING (N.). A passage through a *pillar* of coal.

JUSTICE-MAN (S.). See *Check-weighman*.

## K.

KANK (M.). A twist or snick-snarl in a rope.

KEEKER (N.). An inspector over *hewers* or other workmen underground.

KEEL (N.). (A Saxon word for a long ship). An oval shaped strong and clumsy flat-bottomed vessel for carrying coals from *staithes* or *drops* to ships; about 20 tons capacity.

KEEL-BULLIES (N.). Men who navigate and ply the *puys* of *keels*.

KEEL DEETERS or KEEL DOCTORS (N.). Women and girls who sweep out *keels* and have the sweepings as a perquisite.

KEELERS (N.). See *Keel-bullies*.

KEEPER. (Engine-keeper, Horse-keeper, &c.) See *Brakesman*.

KEEPS or KEPS. See *Cage Shuts*.

KELF (D., Lei.). The vertical height of the back cutting of the *holing* at any time during the operation of *holing* a *stint*.

L

KELVE (I.). See *Bat*.

KENNEL (M.). A collier's term for *cannel*, which see.

KENNER! (N.) An expression meaning *time to leave off working*, conveyed into the workings by shouting, rapping, &c.

KEP. See *Kip*.

KEROSENE SHALE (N. S. W.). Oil-producing shale.

KETCHES (S. W.). See *Backstays*.

KETTLE (S.). A barrel in which men *ride* in a *shaft*.

KEVILS (N.). The weights of coals sent out by the various *hewers* during a certain period.

KEY. A kind of spanner used in boring by hand. Two kinds of *keys* are employed, one for taking the weight off the *rods* (2), at the top of the *borehole* (1), when taking them off or putting them on; it fits the rods, which are lowered back until a box (screw joint enlargement) rests upon it; it usually has an arm on each side to assist in screwing off the rods. The other is an ordinary *key* which is used for screwing and unscrewing the rods as well. See Fig. 86.

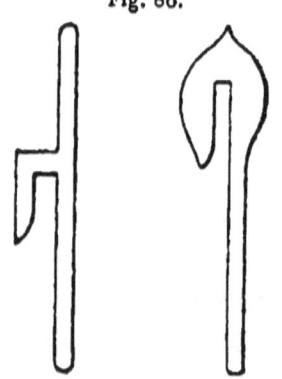

Fig. 86.

KIBBLE. See *Bowk*, but often made with a bow or handle and carrying over a ton of *débris*.

KIBBLES (S. S.). See *Crank*.

KICKER. A liberating catch made in the form of a bell crank lever rocking on a horizontal axis. Used in Kind's system of deep boring.

KICK-UP (N.). See *Tipper*.

KIDING (N.). See *Holing*.

KILKENNY COAL (I.). See *Anthracite*. This Irish coal weighs 99 lbs. per cubic foot.

KILL. To mix atmospheric air with *fire-damp* or other gases so as to make them harmless.

KIND. Generally signifies tender, soft, or easy to work, e. g. a *parting* is said to be *kind* when it allows of an easy separation. *Blue bind* is called *kind blue bind* when it is soft and jointy and easy to *sink* through.

KIND-CHAUDRON (Belg.). A system of *sinking pit-shafts* through water-bearing strata. It consists in boring out the *shaft* from the surface by means of apparatus very similar in kind to that used for *prospective borings*. Not only is the pit bored out but it is lined with metal *tubbing*, and pumped dry without a man ever going down the shaft after the water is met with until it is passed through. The *modus operandi* is somewhat as follows. By means of a very large boring tool a shaft about 5 feet in diameter is first bored out to a certain depth which forms the centre of the pit when fully enlarged. The second operation is to bore out the shaft to the full size with a still larger cutting tool (see *Trépan*) which follows the central hole 10 or 20 yards behind. The *débris* is cleared by means of a large sheet iron *sludger* called a *cuiller*. The boring head is actuated through a lever by steam power, making from eight to ten strokes per minute, and the rate of advance averages about 8 feet per day in ordinary ground. When a suitable stratum has been found upon which to rest the *tubbing*, a watertight ring packed with moss is lowered into position and upon this are built up the rings of *tubbing* placed one upon another at surface, and gradually lowered into the shaft, until the whole of it (in some cases 800 tons) presses and squeezes down upon the moss, forcing it against the sides in such wise as to form a thoroughly watertight joint. The annular space between the rings and side of the pit is filled by means of huge spoons discharged by pistons, with beton or concrete, which when set the water is drawn out of the interior of the pit, and ordinary, or *open-bottom sinking* commenced.

KIND'S PLUG. An ovoid-shaped block of oak fixed to a *boring rod* for jamming into a lining tube of a *borehole* (1) in order to withdraw it.

KINK. See *Kank*.

KIP (N.). A level or gently sloping roadway going *outbye* at the extremity of an engine plane, upon which the full tubs stand ready for being sent up the shaft.

KIRVE (N.). To *hole*. *Kirving* is the same as *holing*.

KIST (N.). A workman's tool box. A cabin in a pit.

KITCHENS. Coal prepared and sold expressly for cooking purposes in ranges, stoves, &c.

KITTY (N.). A length of about 4 inches of straw filled with gunpowder by which flame is communicated to the blasting charge for firing it off in a drill hole.

KNOCK. See *Chap.*

KNOCKINGS (S. W.). Signals made underground by knocking or *jowling* on the coal.

KNOCK OFF. (1.) The point upon an engine plane at which the *set* is disconnected from the rope, or where a *jockey* comes into play.

Fig. 87.

2. A joint for disconnecting the *bucket sword* from the pump rods. See Fig. 87.

*Hoop to keep Joint tight.*

3. To do away with.

4. See *Kenner*.

KNEELER. A quadrant by which the direction of pump rods is reversed.

KÖEPE SYSTEM. *Winding* coals in *shafts* without *drums*, a pulley being fixed upon the main shaft instead. The main *winding rope* has a *cage* at each end, and merely passes half round this *drum pulley*. Under the *cages* ordinary balance or *tail ropes* (2) are suspended. Two additional, or safety ropes, are used, of about one-half the length of the main rope—the cages being attached to each end and small pulleys placed in the *head stocks* carrying them. Fig. 88 is a rough diagram of this system.

Fig. 88.

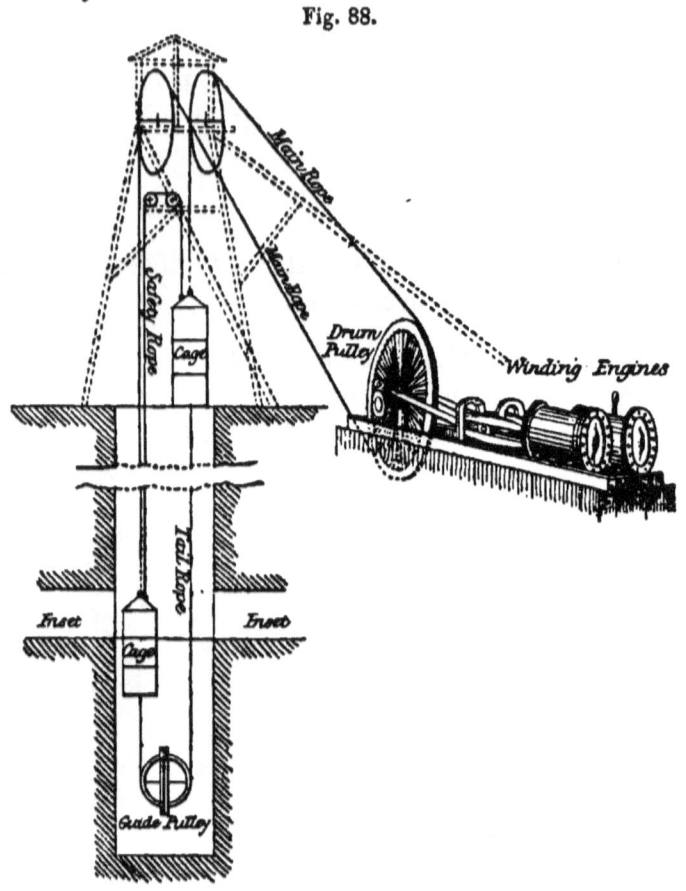

## L.

LACING. 1. (N. S.) Timbers placed across the tops of *bars* or *caps* to secure the *roof* between the *gears*.

2. Strips or light bars of wrought iron bent over at the ends and wedged in tight between the *bars* and the

USED IN COAL MINING, ETC. 151

*roof*, as shown in the sketch Fig. 89. Great elasticity is in this way given to the iron rods, enabling a roof to be very efficiently and economically secured. In

Fig. 89.

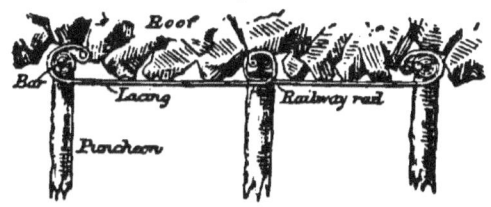

place of wooden bars or head pieces, wrought-iron railway rails are sometimes employed.

LADDERS (Som.). Wooden slides with cross bars placed between them to give steadiness, on which *hudges* run in *steep* seams.

LADE-HOLE (Lei.). A shallow hole cut in the *floor* to receive the drainage, out of which it is laded in buckets.

LAGGING. See *Lacing* (1).

LAGS. Long pieces of timber closely fitted together and fastened to oak *curbs* or rings forming part of a *drum* (3) used in sinking through *quick* (1) ground.

LAID OFF. When operations at a pit are entirely suspended by reason of accident or trade exigencies, the pit is said to be *laid off*.

LAM or LAMB (W.). A kind of *fire-clay*.

LAMB-SKIN (S. W.). See *Culm*. It is sold as such at Swansea.

LAME (F.). The bar to which the cutting teeth of a *trepan* are attached.

LAMESKIRTING (N.). Widening or cutting off coal, &c., from the sides of underground roads in order to give more room.

LAMINGS (N.). Collier's word for accidents of almost every description to men and lads working in or about the mines.

LAMP MEN. Cleaners, repairers, and those who have charge of the *safety lamps* at a colliery.

LAMPS. Signifies *Safety-lamps*, which see.

LAMP STATIONS. Certain fixed places in a mine at which *safety lamps* are allowed to be opened and re-lighted by men appointed for that purpose, or beyond which on no pretence is a *naked light* allowed to be taken.

LAND (F. D.). Rising in the direction of the surface or *outcropping*. *Workings* to the *rise* of a drainage level.

LANDER. The man who receives the loaded *bowk* or *trunk* at the mouth of the *shaft*.

LANDING. A level stage for loading or unloading coals upon.

LANDINGS (S. W.). Coals, &c., sent to *bank*—the *output*, which see.

LANDING SHAFT (S. W.). A *pit shaft* in which coals, &c., are raised.

LANDRY BOX (N.). A wooden spout at the top of a pumping *sett* (1) for carrying off the water delivered by the pumps.

LAND-SALE. The sale of coal, &c., loaded into carts or wagons at the pit's mouth for local consumption.

LAND-SALE COLLIERIES (N.). Those situated in out-of-the-way districts, being unconnected with rail, canal, or sea, and generally working thin or inferior seams.

LAND-WEIGHT (L.). The pressure exerted by the subsidence of the *cover*.

LAP. One coil of rope upon a *drum* or pulley.

LARGE. The largest lumps of coal sent to *bank* (1), or all coal which is hand-picked or does not pass over *screens*, also the largest coals which do pass over *screens*. Lumps weighing upwards of a ton are occasionally sent out at some of the hard or house-coal collieries of Leicestershire.

LAST LIFT (N.). The last *rib* or *jud* to come off a *pillar*.

LATCH. To make an underground survey with a *dial* and chain; or to mark out upon the surface with the same instruments, the position of the *workings* underneath.

LATCHINGS. Diallings or surveys made at a mine.

LATHE! or LAITH! (M.). "Lower the cage down!" or, "Lower down more rope!"

LATHS. See *Lacing* (1).

LAUNDER or LAUNDRY. A wooden or iron cistern or channel in which mine-water is pumped or tipped and conducted away from the pit-top to a water-course or sough.

LAYERED (N.). Choked up with sediment or mud.

LAY OUT (N.). To set out, or put on one side, trams of coals, &c., which have been improperly filled and for which the coal-getters are fined, and the coals in them are forfeited.

LEAD. 1. To haul or draw coals, &c., either by animal or engine power.

2. (Pa.) A stage worked by a mule or by a locomotive engine, of a maximum distance of say three-quarters of a mile.

LEADER. 1. A cast or wrought-iron ring or shoe, bolted to the bottom (often round the outside) of a brick cylinder, a wooden drum, or a wrought-iron cylinder when used for *sinking* through quicksand or gravel. It enables the drum or cylinder to force its way through the ground.

2. (Som.) The *slip* of a *fault*.

3. Any particular or constant bed or band of coal, ironstone, &c., in connection with certain workable beds, serving as a kind of datum line, so to speak, in a mine.

4. (N.) A BACK (1) or fissure in a coal seam.

LEADING BANK (Y.). A breadth of about 18 yards of coal taken out between pairs of *boardgates* to the *rise* commencing from the *bank level*. See Fig. 9 [*Bankwork*].

LEADING MAN. See *First Man*.

LEAN (D.). Thin, poor; of inferior quality.

LEAP. A fault of dislocation or *throw*. There are *Leap-ups* and *Leap-downs*. See *Down-leap* and *Up-leap*.

LEA-STONE (L.). Laminated sandstone.

LEATHER-BED (M.). A tough leather-like clayey substance running in a *fault slip*, composed of the ground-up and squeezed fractured ends of the *coal measures*. Seldom more than a few inches in thickness.

LED (N.). A *led tub* means a spare one, or one which is being loaded whilst another is being emptied.

LEG. 1. (S.) A wooden prop supporting one end of a *bar*.

2. (Y.) (Cleveland.) A stone which has to be wedged out from beneath a larger one.

LEVEL. A road or way running parallel or nearly so with the *strike* of the *seam*, and often used as a *water-level* for drainage purposes.

LEVEL-FREE (W.) Old coal or ironstone *workings* at the *outcrop*, worked by means of a *day level* driven into the hillside.

LEVEL TONS. Weight of mineral *wrought* in tons, any odd cwts. not being taken into account.

LEYS or BLUE-LEYS (L.). See *Bind*.

LID. 1. A short piece of timber about 2 feet long placed atop of a prop to support the *roof*.

2. (F. D.) The *roof* of an *Ironstone working*.

LIDSTONE (F. D.). The roof-stone of an iron mine.

LIE. Having reference to the *dip* of the strata.

LIE-TIME (S.). A period of rest or cessation from work during a *shift* or *turn* (1).

LIFE. When in cutting or getting coal it makes a

crackling or bursting noise and works easily, it is said to have *life* in it, or to be *alive*.

LIFT. 1. The vertical height travelled by the *cage* in a *pit-shaft*.

2. A column or *sett* (1) of pumps.

3. A certain thickness of coal worked in one operation.

4. (N.) To clear *gas* out of a *working place*.

5. To *creep*, as when the *floor* rises up towards the *roof* or *lifts*.

6. A broken *jud* (1).

7. (Pa.) A block of coal measuring three-quarters of a mile on the *strike* by 1000 yards to the *rise*.

8. (F. D.) A rise in the price of coal or in miners' wages.

9. To break up, *bench* (2), or blast coals from the bottom of the seam upwards.

10. A certain vertical thickness of coal seams and *measures*, having considerable inclination, between or in which the *workings* are being carried on to the *rise*, all the coals being raised from one *pit bottom*. A colliery may be composed of several *lifts*. See *Relevée*, Fig. 110.

LIFTING (S.). Drawing *hutches* out of the *working places* into the *main roads*.

LIFTING DOGS. See *Crow's foot*.

LIFTING GUARDS. Fencing placed round the mouth of a *pit-shaft*, which is lifted out of the way for *decking*, by the *cages* as they reach the surface.

LIFTING WICKET (S. W.). See *Lifting guards*.

LIG (N.). To lie down.

LIGNITE. A coal of a woody character, containing about 66 per cent. of carbon, found in the Secondary and Tertiary rocks.

LIGHTNING EXPLOSION. An *explosion* of *firedamp* caused by an electric current during a thunderstorm going into a mine and igniting the *gas*.

LILLYCOCK (M.). See *Kenner*.

LIME CARTRIDGE. A charge or measured quantity of compressed dry caustic lime made up into a *cartridge* (2), and used instead of gunpowder and in a somewhat similar manner for breaking down coal. The cartridge is first placed in the *bore-hole* and *stemmed*, and then water is injected into the hole and on to the lime. Heat or steam is immediately produced, and, expansion taking place, the coal is thereby broken down in a very safe manner, as there is no flame to cause an explosion of gas, and in a less shattered condition than with the use of powder.

LIME COAL. Small coal suitable for lime burning.

LIME PROCESS. The method of getting coal by the use of the *lime cartridge*.

LIMMERS or LIMBERS. Light wooden or iron shafts for attaching pit ponies to the *trams*, especially useful in seams having a considerable inclination.

LINER (Lei.). A *bar* put up between two other bars to assist in carrying the *roof*.

LINES. Pieces of twine about two or three feet in length weighted at the bottom end with a small lump of clay or with a bit of iron, &c., to steady them, and suspended from hooks driven into wooden plugs called *stomps* (which see). Not less than two (called a pair of lines) are put up, their object being to keep the *heading*, &c., in which they may be placed in the proper course or point. A line drawn between the centres of these two strings represents the bearing or point of the compass to be *driven* by, which is determined by the *dial*.

LINING (D.). *Clay Ironstone* in beds or bands.

LINN and WOOL (L.). Streaky grey sandstone.

LINSEED EARTH (Sh.). Blackish grey clay suitable for making into firebricks.

LINSEY (L.). Strong *Bind*, also streaky sandstone.

LIP. 1. (M.) The low part of the *roof* of a *gate-road* near to the *face*; taken down or *ripped*, as it is called, as the *face* advances.

2. The edge of a *fault slip*.

LIPEY BLAES (S.). Lumpy *Bind* or shales.

LIPPEN (N.). To calculate, guess, reckon upon, &c.

LIST. Mine Inspector's term for the schedule of particulars of accidents enumerated in his annual Report to the Government.

LOADER. One who fills the *trams* in the *working places*.

LOADER OFF. A man who regulates the sending out of the full *tubs* from a *long-wall stall, gate end*.

LOADINGS. Pillars of masonry carrying a drum or pulley.

LOAM. Any mixture of sand and clay which is neither distinctly sandy nor clayey.

LOCKER (M.). A short iron or wooden bar for scotching *tram* wheels on inclined *roads*.

LODE (S. S.). A seam or mine.

LODGE. A subterraneous reservoir for the drainage of the mine, made at the pit bottom, in the interior of the workings, or at different levels in the shaft.

LODGMENT (S.). See *Sump* and *Lodge*.

LOFTHEAD (N. S.). A cavity or vacant space in the *roof* produced by a *fall*.

LOFTING. 1. (S. W.) An old or disused *heading* over the top of another one.

2. (N.) See *Lacing*.

LOG (N. S.). See *Dolly*.

LOGGED UP. Supported by *trees, props,* or *puncheons*.

LOOKING (N.S.). Examining the strata which is not *walled* up in a *sinking-pit*.

LONG PAY (S. W.) A system of paying wages.

LOLLEY (M.). See *Locker*.

LONG PILLAR WORK. A system of working coal seams in three separate operations. First, large pillars, one of which is represented by the square *a, b, c, d,* Fig. 90, are formed. Secondly, a number of parallel headings are driven through the block; and, lastly, the ribs or narrow pillars are worked away, commencing in the middle at *e* and working both ways.

Fig. 90.

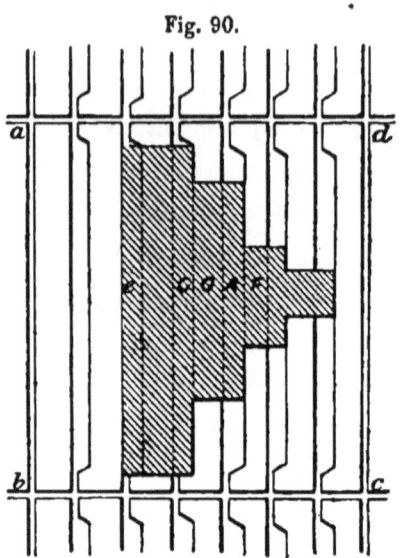

LONG-SHIFT (S.). From say 6 A.M. on Sunday till 6 A.M. on Monday, the time during which the *furnaceman* and horse-tender has to be underground under certain circumstances.

LONG-TON. A weight of more than 20 cwt. In canal trade sometimes 25 or more cwt. of coals are allowed to the ton.

LONGUES TAILLES (F.). See *Long-wall*.

LONG-WALL. A system of working coal and ironstone in which the whole of the seam is *gotten* or worked away, and no *pillars* left in excepting the *shaft pillars*, and sometimes *main road* pillars, the *goaves* being more or less filled up to prevent large accumulations of *fire-damp*. There are two modes of working under the *long-*

*wall* plan. No. 1, to work outwards, commencing near the *shafts* and taking out all the coal, carrying the roads in the *goaf* by *pack walls*; or, secondly (No. 2),

Fig. 91.

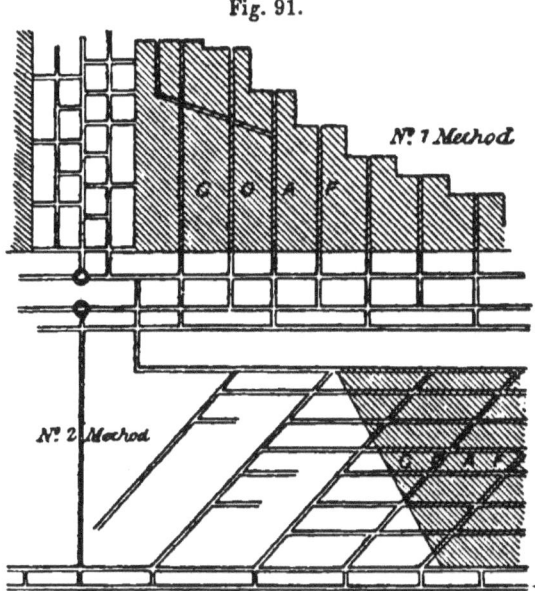

by driving out the *main roads* to the boundary and then *bringing back* the *faces* and leaving all the *goaf* behind.*
See plan, Fig. 91.

* In the *Long-wall system* the *weight* assists greatly in extracting the coal, an advantage lost by other systems of working. See Fig. 92,

Fig. 92.

showing how the subsidence of the *roof* helps to break down the coal at the *face*.

M

LONG-WEIGHT. See *Long-ton.*

LONG-WORK. 1. (Y.) A system of working coal somewhat in the manner shown in Fig. 17.*

2. (Lei.) Ancient plan of working the Main coal of Moira. Each stall or *long-work* was about 150 yards in length (usually two in a pit), and was worked by about twenty *butties*, the coal being got on the *gob-road* system.

LOOKING (N. S.). Examining the unwalled sides of a *sinking pit.*

LOOPS. See *D links.*

LOOSE! or LOOSE ALL! (N.) See *Kenner.*

LOOSE END. The limit of a *stall* next to the *goaf*, or where the adjoining stall is in advance.

LOOSE NEEDLE. See *Dialling.*

LOOSING (S. S.). Lowering a *cage*, &c., into or down a *shaft* or *pit.*

LORDSHIP (S.). Royalty or *acreage rent.*

LORRY (Y.). A running bridge over a *sinking pit* top upon which the *bowk* is placed after it is brought up for emptying.

LOSE. 1. To work a seam of coal, &c., up to where it dies out or is faulted out of sight. This is called *losing the coal.*

2. To be unable to work out a *pillar* on account of *thrust, creep, gob-fire,* &c.

3. A *pit-shaft* is said to be *lost* when it has run in or collapsed beyond recovery.

LOUGHS (L.). Irregular cavities in iron mines.

LOW. 1. (N.) A candle or other *naked light* carried by a miner.

2. (F. D.) Minor channels communicating with *horses*, are termed *lows*.

Low Rope (N.). A piece of rope used as a torch.

Lum. 1. (N.). A chimney placed on the top of an *upcast* shaft to carry off the smoke, &c., and to increase the ventilating current.

2. (D.) A basin or natural swamp in a coal seam, often running several hundred yards in length.

Lumberings (D.). *Bumps* over old *workings*.

Lumps (S. S.). Coal of largest size by one.

Lurry. 1. (Y.) A *tram* to which an *endless rope* is attached, fixed at the *inbye* end of the *plane*, forming part of an appliance for taking up the slack rope. See Fig. 93.

Fig. 93.

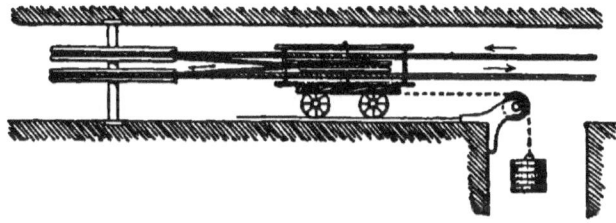

2. A movable platform on wheels, the top of which is made on a level with the *bank* (1) or surface. It is run over the mouth of a *pit-shaft* for a *bowk* to be lowered down upon when reaching the pit top.

Lye (S.). A siding for *tubs* in a mine.

Lypes (S.). Irregularities in the *roof* indicating danger from *falls*.

M 2

## M.

MACHINE. A weighbridge or weighing machine upon which wagons, trams, carts, &c., are weighed, either with or without their loads of coals, &c.

MACHINE-MAN. One who weighs coals, &c., and keeps an account of the number of *tubs* sent to *bank* (1).

MACHINE WALL. The *face* at which a coal-cutting machine works.

MAIDEN FIELD or GROUND. A *coalfield*, &c., which has not been *tapped*.

MAIN DOOR. See *Bearing Door*.

MAIN BOARD-GATE (Y.). The *heading* which is driven to the *rise* of the *shaft*. It is usual to make it larger in sectional area than an ordinary *board-gate*. See *a*, Fig. 9 [Bank-work].

MAIN ENGINE (N.). The surface pumping engine, usually of the Cornish type.

MAIN ROAD. The principal underground *way* in a *district* along which the produce of the mine is conveyed to the *shafts*, generally forming the main *intake air course* of each *district*.

MAIN ROPE. A system of underground *haulage* in which the weight of the empty *tubs* is sufficient to draw the rope *inbye*.

MAIN SUIT (B.). A heavy spring or *feeder* of *water*.

MAINTENAGE (F.). The *face* of workings in *rearing* or vertical seams, consisting of a series of little steps each about six feet in height, and forming the working place of one man.

MAIN SEPARATION DOOR. See *Bearing Door*.

MAKE GAS (M.). A seam of coal which gives off *firedamp* is said to *make gas*.

MAKINGS (N.). The slack and dirt made in *holing*.

MALM (Som.). Loam.

MANAGER. An official who has the daily control and supervision of a colliery or mine, both under and above ground. He usually has the appointment of all the sub-officials employed underground; has the setting out and superintendence of all new works; is responsible to the Owner or Agent for carrying out the requirements of the Act of Parliament, &c.; for keeping up an adequate amount of ventilation; for having the *plans*, books, &c., made and kept up from time to time; and for the general maintenance of order, regularity, and efficiency of everything connected with the *getting* and *output* of the coal, &c. He must hold a Certificate of Competency or of Service from the Government.

MAN HOLE. A *refuge hole* constructed in the side of an underground engine plane or horse road, placed 20 yards apart on engine planes and 50 yards on horseways.

MAN HUDGE (G.). A kind of barrel or box in which men *ride* in a *pit-shaft*.

MAN-O'-WAR (S. S.). A small auxiliary pillar of coal left unworked in the Thick coal-seam workings, as an additional support, or having some special service in regard to *faulty* coal, &c.

MAN ROPE. A *winding rope* used exclusively for lowering and raising men and animals at the time when *tucklers* and *swinging bont* were used and *cages* unknown. When used, the coal-drawing ropes were drawn out of the way up against the shaft sides, and the *man rope* was then swung into the centre of the pit, having its own pulley in the *head gear* fixed between the other two. A separate *drum* (1) was employed for this rope, put into gear when required.

MAN WAY (Pa.). A *bolthole* between two *chutes*.

MARCH (S.). The boundary of the coal or colliery.

MARCHING (S.). Boundary *workings*.

MARCH PLACE (S.). A heading working up to or alongside the *march*.

MARK. Word applied to a band of hemp, &c., wrapped round a *winding rope* to indicate to the *engineman* the position of the load in the *shaft*.

MARL. Indurated clay or shale, sometimes *fire-clay*.

MARROW (N.). A mate, *butty*, or partner.

MARSH GAS. In mining language synonymous with *fire-damp*.

MASSIFS LONGS (F.). *Pillars* in *long-wall* workings.

MASTER CHARGEMAN. The head *sinker* of a *shift*. He prepares and *fires* (2) the *shots*, and looks after the work being properly done, and the safety of the pit and men under him.

MASTERS. Colliers' term for the owners of the works. A pit is said to be worked by the *masters* when the *butty system* is not in vogue. Coals cut by men who are paid by the time and not by the ton or *score* are called *masters' coals*, and are marked or chalked in a particular way in the pit to distinguish them.

MATCH. Gunpowder put into a piece of paper several inches long, and used as a *fuse*.

MATHER AND PLATT'S SYSTEM. Boring or prospecting for coal, &c., by steam machinery with a flat hemp rope instead of rigid rods. The cutters or *boring-head* and rope are raised by a vertical steam cylinder, and have a free fall, varying in height from 2 feet 6 inches upwards. The weight of the cutting tools with guide bar and mechanism for rotating the same is about a ton, but heavier for larger holes. Solid *cores*, showing the *dip* and character of the strata bored through, can be brought to surface, and holes up to 2 and 3 feet in diameter bored to a great depth.

MAUL (N.). A *driver's* hammer.

MAUNDRIL. A *pick* with two shanks and points used in *getting* coal, &c.

MAVIES (N.). Possibly, perhaps.

MEASURE (Sh. S. S.). A bed or *pin* of *ironstone*.

MEASURES. Strata. See *Ground*.

MEASURES HEAD. A *heading* or *drift* made in various strata. See *Crut*.

MEEND or MEAND (F. D.). Old ironstone workings at the *outcrop*, some of which were worked by the Romans.

MEET. To keep pace with: e.g. to keep up the supply of coals at the pit bottom as fast as the *winding engine* can raise them, which is commonly called *meeting the turn*.

MEETING. 1. A siding or pass-by on underground roads.

2. The point in the *shaft* at which the *cages* pass one another or meet.

MEND. To load or reload *trams* at the *gate-ends* out of smaller *trams* used only in the working *faces* in *thin seams*.

MENDITS (F.). See *Putters*.

MENU (Belg.). *Slack*.

METAL (N.). Indurated clay or shale. See *Bind*.

METAL DRIFT (L.). A *heading driven* in *stone*. See *Crut*.

METAL MAN (L.). One who repairs underground roads.

METAL RIDGES (N.). Pillars forming themselves into supports to the *roof*, formed by the *creep* in the *boards*. See Fig. 94.

Fig. 94.

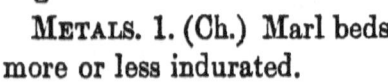

METALS. 1. (Ch.) Marl beds more or less indurated.

2. (S.) Coal seams, or mines of coal, &c.

METAL STONE (N.). Sandstone and shale mixed.

METAL TUBBING. See *Tubbing*.

MIDDLING (L.).

MIDGES (N.). Lamps (not safety) carried by *putters* &c.

MID-WORKINGS (S.). *Workings* with other *workings* above and below in the same *mine* (3) or colliery.

MINE. 1. *Ironstone*, either in thin bands, or in one bed several inches in thickness.

2. A *seam* of coal.

3. A coal-pit or colliery, or a pit or place where *ironstone*, clay, shale, rock-salt, stone, &c., are worked or mined.

4. (S.) A *cross-measures drift* or incline communicating with two or more seams of coal, &c.

5. (S.) A trial *heading* to prove minerals, &c.

MINE EARTH (N. S.). Synonymous with *ironstone* in beds: a term used as much as 200 years ago.

MINE GROUND. Strata containing *ironstone* in layers.

MINE MEASURES (F. D.). See *Mine ground*.

MINERS' COAL TON. In Wales, 21 cwts. of 120 lbs. each.

MINE WORK. An *ironstone mine* (3) or *workings*.

MINGE or MINGY COAL. Coal of a tender nature.

MINGLES (S.). The vertical timbers of the upper part of a pulley frame, on the top of which the pulleys are fixed.

MINIMUM RENT. The certain, dead, or fixed rent payable by the Lessee of a colliery, &c., each half-year, whether he shall have worked or disposed of any minerals or not during that period. The amount payable during the *sinking* of the *shafts* and opening out the underground *workings* is usually less than when the mine has become fully developed.

MISTRESS (N.). A wooden or tin box, having the front open, in which a candle is carried in a pit.

MIZER. The chief tool used in certain systems of *sinking* the cylinders of small *shafts* through water-bearing strata, to remove the ground from beneath them. It consists of an iron cylinder, varying in diameter from 1 foot 6 inches to 6 feet, with an opening on the side and a cutting lip, and which is attached by a box-joint to a set of *boring rods*, and turned from above.

MOBBIES (S. S.).

MONITOR (U. S. A.). See *Gunboat*.

MONKEY (Lei.). An iron catch or *scotch* (1) fixed in the *floor* of a *way*.

MONKEY GANGWAY (Pa.). An *air course* driven parallel with a *gangway* and *heading* at a higher level, and generally in the top-rock or *roof*, and connected with them by *cross cuts*.

MORTS TERRAINS (F.). *Barren* or dead ground. The water-bearing strata overlying the coal measures.

MOSH (Lei.). Synonymous with smash. Coal which is very *nesh* or tender is liable to *mosh down*, or break up into slack, if roughly handled, conveyed long distances, or allowed to stand exposed to the weather for a considerable time. A collier's term only.

MOSS BOX. A cast iron annular open-topped box or ring, placed in watertight ground for making a watertight seat or bed for the *tubbing* of a *Kind-Chaudron* system *sinking pit*. The box is filled with dry moss and is lowered into the pit with, or suspended from, the *tubbing*, the pressure of which, as it settles down, causes compression of the moss to the perfect exclusion of water from behind. It is practically an enormous stuffing-box, and serves the purpose of a *wedging crib*.

MOTE or MOAT. A straw filled with gunpowder for igniting a *shot*.

MOTHER OF COAL. Sooty coal.

MOTHERGATE (N.). A road in the workings to be eventually converted into a *main road*.

MOTIVE COLUMN. The length of column of air in the *downcast shaft* which would be equal in weight to the difference of the weight of the air in *downcast* and *upcast* shafts. The power obtained by *furnace ventilation* is measured by the difference between the weight of the air in the two shafts. To find the *motive column* the following formula is given:—

$$M = D \frac{T-t}{T+459}.$$

M = Motive column.
T = Temperature of upcast.
t = Temperature of downcast.
D = Depth of downcast.

MOTTY (Y.). See *Tally*.

MOUTH. The top of a *pit-shaft* at the surface.

MOUTHING (S. S.). See *Inset*.

MOVE (N. W.). A *roof* which is just about to fall or *weight*.

MUCK (Y.). See *Dirt*.

MUESELER LAMP. A safety lamp brought out and exclusively used in the collieries of Belgium. It is considered the safest lamp of all the many different forms hitherto constructed. Its chief features consist in the horizontal gauze and conical metallic chimney with which it is fitted, making it very sensitive to firedamp, self-extinguishing in an explosive mixture or when not placed perfectly upright, and is a lamp which will withstand a considerable current of air or explosive mixture without going out or causing the flame to pass through the gauze and thereby cause an *explosion*.

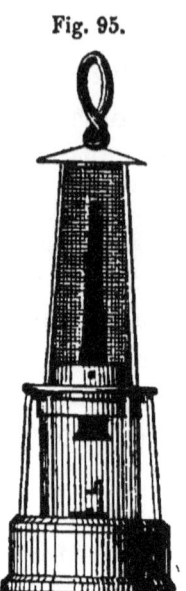

Fig. 95.

MULNIELLO (It.). A kind of quarry or place in a coal mine where stone and débris are obtained for the purpose of *stowing* or filling up *goaves*.

MUSH [rhyming with push] (Lei.). Soft, sooty, dirty, earthy coal, &c.

MUSHY COAL (Lei.). Where a sooty substance pervades coal, or where it is crushed.

MUSSEL BAND. A bed of *clay ironstone* containing fossil bivalve shells, *anthracosia*, &c.

MUTHUNG (Pr.). A concession of mines from the State, generally about 612 acres, described in plan by straight lines and in depth by vertical planes.

# N.

NAGER (B.). A drill for boring holes for *shots*.

NAKED LIGHT. A candle or any form of lamp which is not a *safety lamp*.

NANNIES (Y.). Natural joints, cracks, or *slips* (2) in the *coal measures*. See *Cleat* (1).

NAPPES (Belg.). Water-bearing strata.

NARROWS (N.). Galleries or roadways driven at right angles to *drifts* (4), and not quite so large in area.

NARROW WORK. 1. (Pa.) *Headings, chutes, cross-cuts, gangways*, &c., or the *workings* previous to the removal of the *pillars*.
2. A *working-place* in coal only a few yards in width.
3. See *Deadwork*.
4. A system of working coal in Yorkshire. See plan, Fig. 96.

Fig. 96.

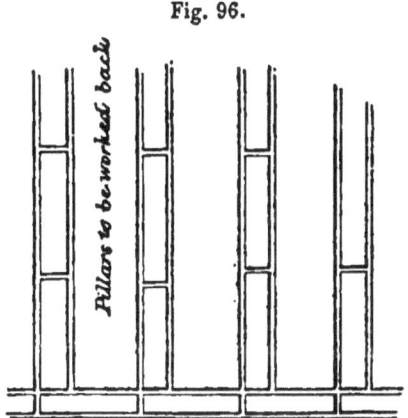

NATTLE (N.). See *Fissle*.

NATURAL VENTILATION. Ventilating a mine without either *furnace* or other artificial means; the heat imparted to the air by the strata, men, animals, and lights in the mine, causing it to flow in one direction, or towards the deepest shaft.

NEEDLE. A sharp-pointed copper or brass rod with

which a small hole is made through the *stemming* to the cartridge in blasting operations.

NESH. Friable, soft, crumbly, powdery, dusty.

NEST-WEISE (F. D.). Iron ore which occurs in pockets is said to lie *Nest-weise*.

NETHER COAL (M.). The lower division of a thick seam of coal.

NICK (N.). To cut or shear coal after *holing*.

NIGHT SHIFT. The set of men who work during the night.

NIGHT WATCH (Lei.). A trusty old collier who keeps guard on the surface during the night.

NIP. 1. (S. W.) A kind of *fault*, the *roof* and *floor* coming nearly together.

2. To cut grooves at the ends of *bars*, to make them fit more evenly.

NIPPING-FORK. A tool formed something like a spanner, for supporting or hanging *boring-rods* at the surface during the screwing on and off of the rods. See *Key*.

NIPPLE (M.). See *Fissle*. A word used to express the crepitant noises made by the settling down or *weighting* of the *roof*.

NITCH WHEELS (S. S.). *Drums* or *pirns* upon which the *wood-chain* winding *bands* coil.

NOG. See *Cog* and *Chock*.

NON-SEAT (M.). See *D Link*.

NOOK (N.). A corner of a *pillar* of coal.

NOOPER (Lei.). A *Dresser*, which see.

NORTH END (Y.). The *rise* side of the coal in North Yorkshire.

NOSE IN. A stratum is said to *nose in* when it *dips* beneath the ground or into a hill-side in a V or nose form.

NOSE OUT. A nose-shape stratum *cropping* out.

NOTCH STICKS (F. D.). Short pieces of stick notched or nicked, used by miners as records of the number of *tubs* of coal, &c., they send out of the pit during the day.

NUBBER (M.). A block of wood about twelve inches square, for throwing *tubs* off the road in case the couplings or ropes break. A boy places it between the rails as soon as the full train has passed *outbye*.

NUTS. Small lumps of coal which will pass through a *screen* the bars of which vary in width apart between ½ inch and 2½ inches.

## O.

OBERBERGAMT (Pr.). A board or council consisting of six or seven members, which sanctions colliery rules, prescribes as to the duties of *inspectors, fiery mines, safety lamps*, &c. The State has appointed five mining boards, or *Oberbergämter*.

OBERSTEIGER (Pr.). An underground *overman*, who acts under the guidance of the *Betriebsführer*, or *manager*.

OCEAN COAL (C.). Coal-seams lying beneath the sea.

OCHRE. See *Canker*.

ODD-KNOBBING (S. S.). Breaking off the coal from the *sides* in the Thick-coal *workings*.

ODD MAN. One who works by time at sundry jobs in the mine.

ODD WORK. Work other than that done by contract, such as repairing roads, constructing *stoppings, dams*, &c.

OFF (N.). Worked out, *gotten, wrought*.

OFF-GATES (N.). *Goaf* roadways in *long-wall workings* about 120 yards apart.

OFF-TAKE. 1. The raised portion of an *upcast* shaft above the surface, for carrying off smoke and steam, &c., produced by the *furnaces* and engines underground.

2. The length of *boring-rods* unscrewed and taken off at the top of the *bore-hole* (1), depending upon the height of the *head-gear* and depth of the *staple*, or well.

OFF-TAKE RODS. Auxiliary wooden rods at the top and bottom of a *winding-shaft*, by means of which the *cages* are guided and steadied during *decking*.

OIL-SHALE. Shale containing such a proportion of hydrocarbons as to be capable of yielding mineral oil on slow distillation. Occurs in layers or seams interstratified with other aqueous deposits, as in the Scottish coal-fields. It consists of fissile argillaceous layers, highly impregnated with bituminous matter, passing on one side into common shale, on the other into *cannel* or *parrot* coal. The richer varieties yield from 30 to 40 gallons of crude oil to the ton of shale.

OLD MEN. The former workers of a mine. The workings left by them are called *old men's workings*, or, as in Derbyshire, *The old man*.

ON-COST (S.). *Dead work* expenses, being costs incurred at a mine, whether minerals are raised or not.

ONE WAY (S. S.). A particular class of *house* coal.

ON-SETTER. See *Bottomer*. Also the man who changes the *tubs* in the *cages* at *bank* (2).

ON-SETTING MACHINE. A mechanical apparatus fixed at the top and at the bottom (or only at the surface) of a *pit-shaft*, on a level with the *cages*, for loading them with the full *tubs*, and discharging the *empties*, or *vice versa*, at one operation, thus effecting a great saving of time and manual labour. There are several machines for performing this important operation, viz., Fowler's hydraulic apparatus, by which cages having three or four *decks* can be loaded and unloaded in a few seconds without moving the *winding-engine* or *decking*, as it is called, in the ordinary sense of the word. Another machine takes the form of an inclined framework, carrying the *tubs*, which the *cage* actuates on being lowered on to the *props* or *keeps*. A third is worked by a small steam cylinder, which tilts a platform carrying the *trams*, thus causing them to run forward on to the *cage*. A fourth consists in withdrawing the full *trams* from the *cages* by means of a light rod and a chain worked by a small steam-engine fixed near the top of the *screens*, which are directly opposite the *pit-top*, thereby avoiding almost all the heavy work of pushing heavily loaded *trams* about on surface, which occasionally carry 25 cwt. of coals, the *tram* being 9 cwt.

ON THE RUN (Pa.). The ability to work a seam of coal which has sufficient inclination to cause the coal, as worked away towards the *rise*, to fall by gravity to the *gangways* for loading up into *cars*, is called working coal *on the run*.

OPEN BOTTOM. The bottom of a *sinking-pit* open directly to the atmosphere or surface.

OPEN-CAST WORKING (S.). A *coal-working* having no *roof*. See *Open Hole*.

OPEN HOLE. Coal or other mine *workings* at the surface or *outcrop*, sometimes carried to a depth of 50 or 60 feet, forming a kind of quarry. See *Bench Working* (Fig. 15).

OPENINGS. 1. Short *heads* (1) driven at certain intervals between two or more parallel *heads* or levels for ventilation. As each *opening* is cut, the last one is built up with bricks and mortar, to drive the air-current forward to the *face* (1) of working.

2. (N.) *Backs* (1).

OPEN LIGHT. See *Naked Light*.

OPEN OFF. To commence the working away of a seam of coal, &c., upon the *long-wall* system from the *shaft pillar*, or it may be the far end of the *royalty* (1), or from any *headings* previously driven out for the purpose of commencing such system, or a modification thereof.

OPEN OUT. To drive *headings* out, or commence working in the coal, &c., after *sinking* the *shafts*.

OPEN ROCK. Any stratum capable of holding much water, or conveying it along its bed by virtue of its porous or open character.

OPEN SHELL-AUGER. A coal-boring tool for extracting clay and other débris from the hole: it has no valve at the lower end.

OPEN-TOP TUBBING. A length of *tubbing* having no *wedging-crib* on the top of it.

OPEN WORKINGS. *Workings* carried on by *open hole.*

OUTBREAK COAL. An old term for *outcrop* of a coal seam.

OUTBURST. 1. (N.) A *Blower.*
2. See *Crop.*

OUTBYE. In the direction of the pit bottom.

OUT-CROP. 1. The surface-edge of any inclined stratum.
2. To incline upwards, so as to appear at the surface.

OUT-DOOR STROKE. That stroke of a Cornish pumping-engine by which the water is forced upwards by the weight of the descending pump-rods, &c.

OUT-FALL. A seam cropping out at a lower level.

OUT-OVER. See *Outbye.*

OUT-PUT. The quantity of coal, &c., raised during a certain period—for instance, 6000 tons per week.

OUT-SET. 1. (N.) The *walling* of *shafts* built up above the original ground-level.
2. A brick or stone *shaft walling* built up within *tubbing.*

OUT-STROKE. The privilege of breaking a *barrier*, and working and conveying underground the coal from an adjoining royalty.

OUTSTROKE RENT. Payment made for the privilege of working through a *barrier*, &c., and conveying the produce of the mine from an adjoining property.

OVERBURDEN. *Cover* in *open workings.* See *Baring.*

OVERCAST. See *Air-crossing*.

OVER-CROSSING. See *Air-crossing*.

OVERGATE. See *Air-crossing*.

OVERGETTINGS. Minerals worked and sold from a *royalty* in excess of the certain quantity upon which a rent or *royalty* at per acre is paid.

OVERHAND STOPING. A system of working thick seams of coal in Germany. The upper divisions are *wrought* first and then the lower. The word *stoping* is one having special reference to metalliferous mining, and not to coal.

OVERLAP FAULT. A peculiar kind of fault where a *seam* is reversed or doubled back over itself. See Fig. 70 (5).

OVERLIE (Som.). The Triassic or other later formation of strata overlying the *coal measures*.

OVERLYING. Rock beds having no true connection with the *coal measures*, but which have been deposited at a subsequent date: e. g., some of the *traps* of the South Staffordshire and Shropshire coal-fields.

OVERMAN, also OVERSMAN. One who has charge of the *workings* whilst the men are in the pit. He gets his orders from the *underviewer*.

OVER-ROPE. The *winding rope* which passes from the pulley over the top of the *drum* (1).

OVERTHROW. 1. (Pa.) Wooden air pipes for connecting *headings* for *ventilation*.

2. (Y.) See *Air-crossing*.

OVER-VENTILATION. Too much *air* in the *workings*.

USED IN COAL MINING, ETC.      181

OVER-WIND. To draw a *cage* or *bowk* up into the *headstocks*.

OXTER (S.).

## P.

PACK. A rough wall or block of coal or stone built up to support the *roof*. Fig. 97.

PACK BUILDER. One who builds *packs*.

PACKER. A man who builds or constructs *packs*.

Fig. 97.

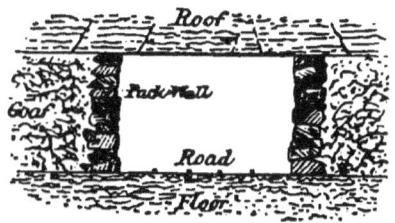

PACK WALL. A wall of stone or rubbish built on either side a *gate road*, to carry the *roof* and keep the sides up. See Fig. 97.

PADDY (Y.). An open or non-safety lamp carried by men and lads in the mines.

PADDY PAN (Lei.). *Skeps* formerly used in *swinging bont*.

PAIR OF GEARS (N.). See *Gears*.

PAIR OF TIMBERS (S. W.). See *Gears*.

PAIRS (S. S.). Two *pit-shafts* sunk to the Thick coal seam about 100 yards apart.

PAN. 1. (Som.) Fire or underclay of the Radstock coal seams.

2. (M.) Sheet-iron vessels holding, say, ¼-cwt., into which *fillers* rake the *small*.

PANE (S. S.). A *lift* (3) or *stint* of coal measuring 2 feet 6 inches high, 6 feet in width, and 6 feet under or forward.

PANEL. A large rectangular block or *pillar* of coal, measuring, say, 130 by 100 yards.

PANEL WORKING. A system of working coal seams which came into use about 1810 in the North of England. See Fig. 98. The colliery is divided up into large squares or *panels*, isolated or surrounded by solid ribs of coal, in each of which a separate set of *boards* and *pillars* is worked, and the ventilation is kept distinct—that is, every panel has its own *intake* and return, the air of one not passing into the adjoining one, but being carried direct to the *upcast* shaft.

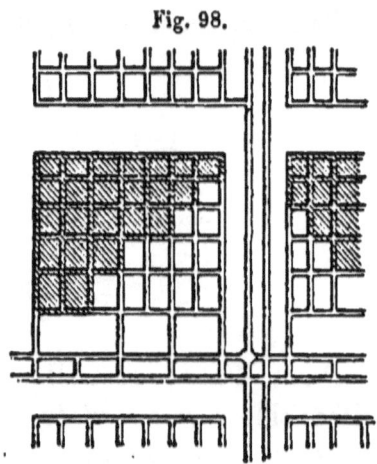

Fig. 98.

PAPER COAL. Finely laminated coal of the Tertiary era, resembling highly compressed leaves.

PARACHUTE. 1. A thin leather washer placed between two stops on the lower end of *boring-rods*, to break the fall of the rods in case they are accidentally dropped or break, by preventing the water in the borehole getting past it beyond a certain velocity. It acts as a kind of cushion or brake.

2. (F.) A safety *cage* fitted up with an ingenious arrangement by which, on the breaking of the *winding-rope*, a wedge is, by the action of springs, inserted between the wooden guides and a part of the cage, so as to bring the latter immediately to a standstill.

PARCEL (S. S.). An old term for a ton; really 27 cwts.

PARROT COAL (S., N.). A description of *cannel coal*, so called because when on the fire it splits and cracks up with a chattering noise, like a parrot talking.

PART CANDLES. The use of candles as well as *safety lamps* in a mine.

PARTING. 1. (S. W.) The double *roads* (2) laid in an *inset* or *pit-bottom arching*.

2. Any thin interstratified bed of earthy material.

PASS-BY. A siding in which *tubs* pass one another underground. In Fig. 99 is shown a plan of a *pass-by* as sometimes constructed upon a self-acting inclined plane.

Fig. 99.

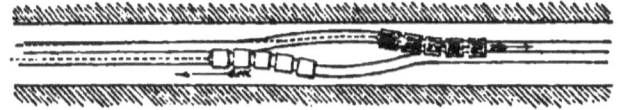

PASS-PIPE. An iron pipe connecting the water at the back of one set of *tubbing* with that of another, or a pipe only in communication with one *tub*, and open to the interior of the shaft.

PATCHING (S. W.). Workings carried on at the *outcrop* or by *open hole*, their depth and extent being limited by the quantity of water met with and the amount of *baring* required.

PATCHWORK (D.). Synonymous with *Patching*.

PATENT FUEL. Small coal, with an admixture of from 8 to 10 per cent. of pitch or tar, compressed by machinery into bricks or blocks of a convenient size for use in the furnaces of boilers, &c.

PAVEMENT. 1. (S.) The *floor* of a mine.
2. (S.) A kind of *fireclay*, *clunch*, &c.

PAY. The day upon which, or the place where, wages are made up or paid. Going to draw wages is called "going to the *pay*."

PEACOCK COAL (L.). Iridescent coal.

PEAT COAL. A soft earthy variety of coal, of Secondary or Tertiary era.

PEAS. Small coals about ½-inch or ¾-inch cube.

PECK. See *Pick*.

PECKING UP (S. S.). Elevating or propping up with rough stones, bricks, rubbish, &c.

PEGGY (Y.). Synonymous with *pick*, which see.

PEGS (F. D.). See *Notchsticks*.

PELDON (S. S.). Hard and compact siliceous rock. See *Cank*.

PELDRIN (N. S.).

PENITENT (F.). A *fireman* who, in early coal mining days, was employed to explode (purposely, in order to get rid of it) the *fire-damp*. So called on account of the resemblance of his dress to that of certain religious orders in the Roman Catholic Church.

PENNYSTONES. Bands of *clay ironstone*.

PENTHOUSE or PENTHUS. A wooden hut or covering for the protection of *sinkers* in a pit bottom.

PFEILERBAU (Pr.). See *Board* and *Pillar*.

PICK. 1. A tool for *cutting* and *holing* coal, generally weighing about 5 lbs. Fig. 100 shows several kinds.

Fig. 100.

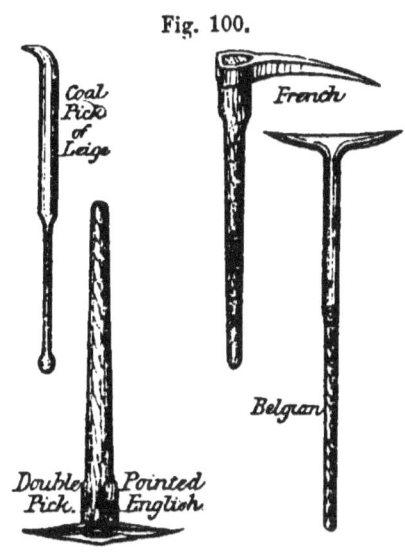

2. To dress with a *pick* the sides of a *shaft* or other excavation.

3. To remove shale, dirt, &c., from coals.

PICK AWAY (M.). To *dip* rapidly.

PICKED BEAT (D.).

PICKER. 1. A sharp-pointed cutting tool used as an accessory to a *mizer*. It is fixed upon the same rods and above the mizer, and indicates the exact position of the latter when in operation.

2. (S.) See *Pricker* (3).

PICKMAN (S. S.). See *Hewer*.

PICK-UP (M.). To reduce the *stock*, which see.

PICK-UPS (M.). See *Tipper*.

PICKWORK. Cutting coal with a *pick*. *Heading* is chiefly done by it.

PIECE (S.). See *Bait*.

PIER-STONE (S.). A very hard variety of freestone.

PIKE. See *Pick* (1).

PIKEMAN. See *Hewer*.

PILING. Driving down into *quick* ground iron-shod 3-inch battens of 12 feet or 14 feet in length, supported by *curbs*, and forming a circle larger than the ultimate size of the *shaft* when *walled* up within. Fig. 101.

Fig. 101.

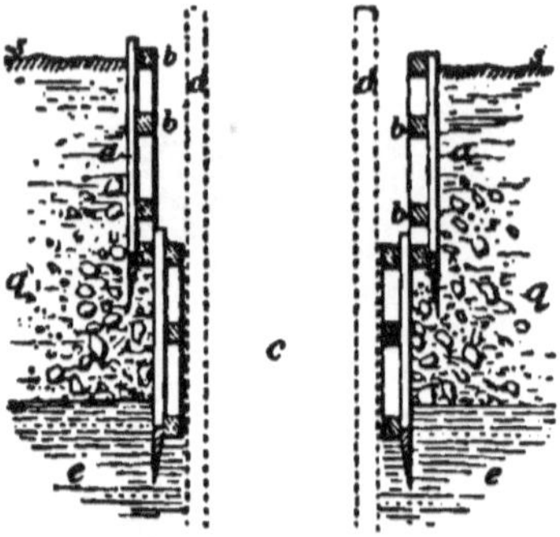

*a a*, Battens. *b b*, Curbs. *c*, Shaft. *d d*, Walling. *s s*, Surface line. *q q*, Quick ground. *e e*, Solid stratum.

PILLAR. A solid block of coal, &c., varying in area from a few square yards to several acres.

PILLAR AND STALL. A system of working coal and other minerals where the first stage of excavation is accomplished with the *roof* sustained by coal, &c. Fig. 102 shows in plan one of the many various modes of working in this manner.

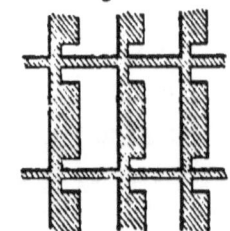

Fig. 102.

PILLARING BACK (N. S.). See *Drifting Back*.

PILLAR MAN (I.). A man who builds stone *packs* in the *workings*.

PILLAR ROADS. Working-roads or inclines in *pillars* having a range of *long-wall faces* on either side.

PILLAR WORKING. Working coal on much the same plans as *Long-pillar* and *Pillar and Stall* systems.

PIMPLEY (Sh.). *Bind* (1) containing ironstone nodules.

PINCH. A kind of crowbar used for breaking down coal, &c.

PIN-CRACKS (Lei.). Small fissures in coal seams filled with *water* and *gas*.

PINDY (I.). See *Kelve*. A term used in the South of Ireland.

PINNINGS (N. S.). *Bratticing* in *headings*.

PINS. Thin beds of *ironstone* of the *coal measures*.

PIPED AIR. *Ventilation* carried into the *working places* in pipes. See *Brattice*.

PIPER (L.). A *feeder* of *gas*.

PIPES. See *Coal Pipes*.

PIRNS (S.). Flat-rope *winding* (1) *drums* (1).

PIT. 1. A *colliery*, a *pit-shaft*, a shallow hole, &c.

2. The *workings*, inclusive of all roads, &c., situated underground.

PIT BANK. The raised ground or platforms upon which the coals are sorted and *screened* at surface.

PIT BARRING (S.). Timbers supporting the sides of a *shaft*.

PIT BOTTOM. The *inset* and underground roads, &c., in the immediate vicinity of the *shafts*.

PIT-BOTTOM STOOP (S.). A large solid block or

*pillar* of mines left ungotten around and in support of the *pit-shafts*.

PIT BROW (L.). See *Pit Bank*.

PITCH. *Dip* or *rise* of a seam.

PITCHER BRASSES (Sh.). Indurated schistose clay.

PITCHERS (N.). *Loaders* in the *pit* (2), and men who take up and relay the rails in the *workings* and *long-wall faces*.

PIT COAL. Generally signifies the bituminous varieties of coal.

PIT-EYE. *Pit bottom*, or the entrance into a *shaft*.

PIT-GATE (Y.). Any place in the immediate neighbourhood of a colliery at which colliers hold meetings of their own in reference to wages, &c.

PIT-HEAD MAN. The *banksman* who has charge of the *pit-top*.

PIT HEAP. See *Heapstead*.

PIT HILL. See *Pit Bank*.

PIT LOG (S. S.).

PITMAN. A *collier* (1); also one who looks after pumps, &c.

PIT-PROP. A piece of fir timber, being part of the stem of a tree, varying in length according to the height of the *workings*, and about one inch in diameter for every foot in length: used as a temporary support for the *roof*.

PIT RAILS. Iron or steel railway rails upon which *trams* or *tubs* run in a mine.

PIT-ROOM. The extent of the underground *workings* in use or available for use.

PIT ROPE. *Winding rope.*

PITS (S. W.). Long open-air fires for converting coal into rude coke for blast-furnace purposes.

PIT-SHAFT. See *Shaft.*

PITTER. A horse or pony suitable for underground work.

PIT-TIP. A bank or heap upon which rubbish out of the mine is tipped.

PIT-TOP. The mouth of a *pit-shaft.*

PIT WOOD. The timber used for propping the *roof,* &c.

PIT WORK. The whole system of pumps and pump-rods, &c., in a pumping or *engine-pit.*

PLACE. 1. A *working place,* or a point at which the cutting of coal, &c., is being carried on.

2. A kind of cabin in which tools, &c., are kept in the mine, and in which a *deputy* gets his *bait* or *snap.*

PLAN. 1. The system upon which a mine is worked, e. g. *long-wall.*

2. A map or plan of the underground *workings,* which in Great Britain must be drawn to a scale of not less than 44 yards to an inch, and must show the whole of the *workings,* accurately marked thereon, at least every six months. The term *plan* also includes a *section* of the mines and of the underground works.

PLANE. A *main road,* either level or inclined, along which coals, &c., are conveyed by engine-power or by gravity.

PLANE BACKS (S.). See *Back* (1).

PLANK (S. W.). Strata drained of *gas*.

PLANK DAM. A watertight *stopping* fixed in a *heading*, constructed of balks of fir placed across the passage, one upon another, sideways, and tightly wedged.

PLANK TUBBING. Shaft lining of wooden planks driven down vertically behind wooden *cribs* all round the shaft, all joints being tightly wedged to keep back the water. See Fig. 101.

PLANT. The *shafts*, engine-houses, railways, machinery, workshops, &c., of a colliery or other mine.

PLASTER (D., N.S., &c.). Gypsum. A fine granular to compact, sometimes fibrous or sparry aggregate of the mineral gypsum. Normally white, but may be coloured grey, brown, yellow, or red. It occurs in beds, lenticular intercalations and strings usually associated with beds of red marl or clay.

PLASTER-PIT (D., &c.). A mine in which gypsum is worked. The system of working is usually a rough kind of *pillar working*, the *pillars* being left sufficiently large to keep the surface from falling in. *Plaster* is often worked by *open hole*.

PLATE. See *Bind* (1).

PLAY. 1. Signifies *not at work* or *standing*.

2. (N.) To work a *steel mill*.

3. Idle—not at work on account of idleness, or for some other particular cause.

PLAY DAY. A day on which, on account of shortness of trade, from accident, or from other causes, minerals are not worked and raised.

PLAYER. A man who used to work a *steel mill.*

PLAY-IN (Lei.). To commence *holing* and *getting* a *face* of coal out of the side of a *heading.*

PLENUM. A mode of ventilating a mine or a *heading* by forcing fresh air into it.

PLIES (S.). Layers of coal or other rocks.

PLUG BOX. A wooden water-pipe used in *coffering.* See Fig. 103.

PLUGGED CRIB (Y.). A *walling crib* carried by iron plugs (two to each segment) fixed in the rock two or three feet in depth.

PLUGGING. Supporting a *crib* upon iron bars fixed in a *shaft side.*

PLUGMAN. An old term for *engineman.*

Fig. 103.

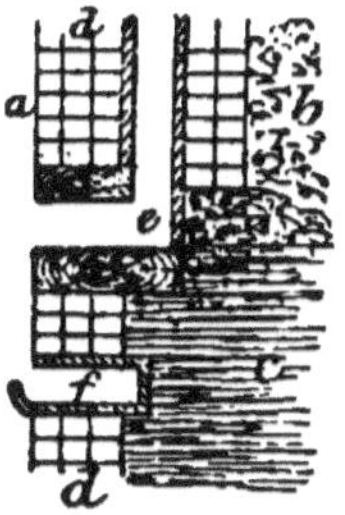

*a*, Shaft side. *b*, Water-bearing ground. *c*, Solid ground. *d*, Walling of shaft. *e*, Plug box. *f*, Water crib or "garland" (1).

PLUM-BULKING (S.). The full *dip* of the coal seam.

PLUM HATCHING (S.). The full *rise* of a coal bed.

PLUM PITCH (B.). The full *rise* or full *dip* of the strata.

PLUMB END (Y.). See *End.*

PLUMP FAIR (S. S.).

PLUNGER CASE. The barrel or cylinder in which a solid piston or plunger works in a *forcing sett* (1) of pumps.

PLUNGER POLE. The solid ram working up and down within a *plunger case.*

PLY (S. S.). A thin bed or band of shale, &c., lying immediately over a coal seam.

POCKET. 1. See *Bag*.
2. See *Swelly*.

POINT. The bearing or direction, in reference to the magnetic meridian, in which an underground road is driven. See *Driving by Lines*.

POLE CASE. See *Plunger Case*.

POLL (S. W.). To clean the shale, &c., off *ironstone*, ready for weighing into *stock*.

PONY-PUTTER (N.). A boy who drives a pony in the *workings*. He is paid at per *score*, *put* 200 yards.

POPPET-HEAD. A shallow pit *pulley-frame*.

PORCH (Y.). The *arching* at the *pit bottom inset*.

PORTEUR (F.). See *Hurrier*.

POST. 1. (N.) A solid block or *pillar* of coal.
2. (N.) Sandstone (fine grained).

POST AND STALL (Y.). A system of working a coal seam much the same as *pillar and stall*.

POSTING (Y.). Extracting the *posts* (1) or working the *broken*. See Fig. 104.

Fig. 104.

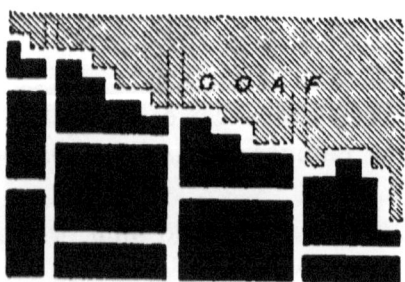

POSTING-HOLE (Y.). See *Bolt*.

POST-STONE. Sandstone rather fine grained.

POT-BOTTOMS (S.). See *Bell-moulds*.

POT HOLE (L.). A small temporary *lodge* in a *sinking-pit*.

POT HOLES. See *Pot-bottoms*.

POT MIZER. A boring tool occasionally used in clays mixed with pebbles. It is made in the form of a spiral cone, which is open at the top to receive the pebbles carried up by the worm on the outside and falling over the edge into the cone.

POUND. 1. An underground reservoir of water. See *Lodge*.

2. A large natural fissure or cavity in the strata.

Fig. 105.

POUNDSTONE (Sh.). A kind of *underclay*.

POUNSON (N. W.). Dense soft clay underlying coal beds.

POUT (N.). A tool used by *deputies* for knocking out or *drawing* timbers in the *workings*.

POXON ROCK (Lei.). A red gravelly stratum (Permian ?) overlying *coal measures*.

PRICKER. 1. A thin brass rod for making a hole in the *stemming* when blasting, for the insertion of a *fuze* or *touch*, and through which aperture the flame obtains access to the cartridge.

2. (S. S.) A long iron rod or poker used for loosening and bringing down the coals from overhead in the Thick coal workings. See Fig. 105.

3. A piece of bent wire by which the size of the flame of a *safety lamp* is regulated, without removing

o

the top of the lamp. It passes up into the lamp through the oil reservoir in a tube.

PRICKING (Lei.). Soft coal or earth for *holing* in.

PRIZE (Lei.). To lift or loosen with a lever or a *pick*.

Fig. 106.

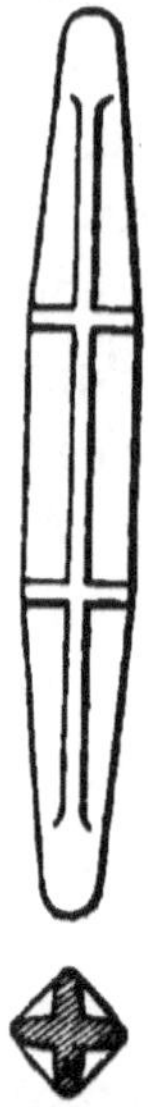

PROP. A wooden or cast-iron temporary support for the *roof*, reaching from the *floor*. When of timber they are generally used of as many inches in diameter as they are feet in length. Fig. 106 shows a cast-iron prop. They are not much used.

PROPPING. The *timbering* of a mine.

PROPS. See *Keeps*.

PROP-WOOD. Timber suitable for cutting, or already cut into *props*. See *Prop*.

PROSPECTING. Examining (by *boring*, sinking trial pits, &c., and geologically surveying) a tract of country in search of minerals.

PROTECTOR LAMP. A safety lamp the flame of which it is impossible to expose to the outward atmosphere, as the fact of unlocking or rather unscrewing it extinguishes the light. (A Mr. Teale of Manchester was the inventor of this self-extinguishing appliance.)

PROUD COAL (S.). That which naturally splits off in flakes or slabs when worked in a particular manner, producing waste and deterioration.

PROVE. 1. To ascertain by *boring*, *driving*, &c., the position and character of a coal *seam*, a *fault*, &c.

2. (S.) To examine a mine in search of *fire-damp*, &c., known as *proving the pit*.

PUCKING or PUCKS (S. W.). See *Creep*.

PUDDING ROCK (Y.). Conglomerate or breccia.

PUDLOCKS. Cross timbers resting upon *horse-trees* against which rubbing-boards work.

PUISARD (F.). See *Sump*.

PUITS (F.). *Shafts* or *pits*.

PULL. 1. To subside or settle down. See *Creep*.
2. The *drag* in *ventilation* of mines.

PULLER-OFF (M.). A man who takes the loaded *trams* off the *cages*, or who withdraws the *empties* from them at the *bottom*.

PULLEY. The wheel over which a *winding rope* passes at the top of the *head-gear*.

PULLEY FRAME. See *Head-gear*.

PULLEYING. *Overwinding* or drawing up a *cage* or *kibble* into the *pulley-frame*.

PULLING BACK. See *Posting*.

PULLING-OVER ROPE. A short light hemp rope for drawing the ends of *winding ropes* over the pulleys off the *drum* (1).

PUMP FIST. The lower end of a *plunger case*.

PUMPING. The operation of filling a *sludge pump* by an up-and-down motion of the rods or rope, called *pumping the sludger*.

PUMP-STOCKS (L.). See *Pump-trees*.

PUMP-TREES. Cast (wrought iron were formerly often used) iron pipes, generally nine feet in length, of which the *column* or *sett* (1) is formed, conveying the water from the pump up the *shaft*. They run up to say thirty inches in diameter, and are bolted together and steadied by *chogs*. Fig. 40.

PUNCH (N.). See *Pout*.

PUNCH AND THIRL (S. S.). A kind of *pillar and stall* system of coal-getting.

PUNCHEON (M.). See *Prop*.

PUNCH PROP (N.). A short timber *prop* set on the top of a *crowntree* or used in *holing* as a *sprag*.

PUT. 1. To haul coal, &c., underground.

2. (Som.) A *box* of a capacity of from 3 to 6 cwt. of coal, used in *thin seams*.

PUTTER. See *Haulier*. Age from 15 to 20 years; paid by the *score* of *tubs*, *put*, say 100 yards. *Putters'* places are *cavilled* for.

PUTTING. See *Haulage*.

PUYS (N.). Great oars by which *keels* are pulled and steered about.

PUT TO STAND (S. S.). Stoppage of coal drawing on account of *firestink*.

## Q.

QUAR or CLIFF QUAR (F. D.). A kind of *Bind* (1).

QUARLS (N.). Fire-bricks.

QUARRY. An underground excavation formed in the *roof* stone or shale or in a *fault*, for the purpose of obtaining material for *stowage* or *pack-walls*. A plan only followed when it is less costly than to leave coal in the mine, or to bring material from surface for such purposes. Fig. 107 is a vertical section.

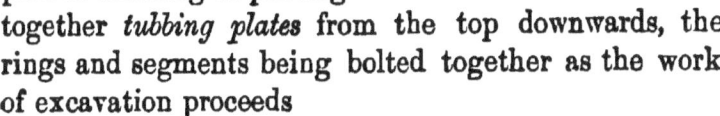

Fig. 107.

QUARTER COAL (Y.). See *Colliers' Coals*.

QUARTERING-IN (L.). A plan of building or putting together *tubbing plates* from the top downwards, the rings and segments being bolted together as the work of excavation proceeds

QUICK. 1. Soft watery strata, such as running sand.

2. (S. S.) Solid or ungotten coal forming the *roof* of a roadway in a Thick coal colliery.

3. Blasting powder is said to be *quick* when it burns or goes off very rapidly.

QUOICENECK (Sh.). Greyish black clay with shining surfaces, and streaked.

R.

RACE. 1. (S.) See *Journey*.

2. The space in which a *drum* (1) revolves.

RADDLE (Y.). Earthy Hæmatite occurring in the *coal measures*.

RAFF-YARD (N.). A walled-in yard on the surface, in which the smiths, wrights, carpenters, &c., work.

RAG AND CHAIN PUMP. One of the earliest contrivances for draining coal pits, consisting of a tube or pipe in which a chain, to which bunches of rags were at intervals attached, was caused by manual labour to carry up water in much the same way as our nineteenth century chain pumps do. These pumps were in use 250 years ago.

RAILS. The iron or steel portion of the permanent or temporary *way* (2). They weigh from 15 to 35 lbs. per yard run; are usually from 6 to 15 feet in length; are either of ⬬ ⬮ ⬯ or ⬰ section; are laid with a gauge of from 1 foot 8 inches to 2 feet 6 inches. Main *engine plane* rails are generally fished. Angle iron ⬱ rails are still in use, but are rapidly disappearing.

RAIN (M.). An underground place is said to *rain* when water drops freely from the *roof*.

RAISE. To *wind* (3) coals, &c., to the surface.

RAISINGS (F. D.). See *Get* (2).

RAIT or RATE (M.). To split off. Coal roads, &c., are said to *rait themselves* when the sides keep splitting or peeling off. Roads driven on the *end* are more liable to *rait* than when driven *face on*.

RAKE (D.). A series of *pins* of *clay ironstone* lying within a few feet or yards of one another in a seam of *bind*, making a *workable ironstone*.

RAKE. 1. (M.) To smother a ventilating *furnace* with fuel, so that it smoulders for many hours, and allows the *upcast* shaft to cool, for the purpose of doing repairs therein, or for other special purposes.

2. (M.) An iron rake with a short handle, with which *fillers* fill *baskets* or *pans*.

RAKERS. Shots placed round *sumpers*.

RAKING-COAL. A large lump of hard coal placed upon a fire or ventilating *furnace*, for the purpose of just keeping it burning, or rather smouldering, when a larger fire is not required.

RAKING PROPS. Short wooden props used in *sinking* for supporting the *curbs* during the excavation of the sides of the *shaft*.

RAM. See *Plunger Pole*.

RAMBLE. See *Falling*.

RAMMELLY (M.). Mixed argillaceous and sandy rocks.

RANCE (S.). A *pillar* of coal—a rarge *stoop*. See *Room and Rance*.

RANDID STONE (C.).

RAP (S. W.). See *Bump*.

RAP IN (Som.). To wedge down blocks of stone in underground quarries.

RAPPER. 1. A lever with a hammer attached at one end, fixed at the pit top or top of an inclined plane, by

Fig. 108.

which signals are given to and from *banksman* or *engineman*. See Fig. 108.

2. (M.) The upper end of the vertical arm of a *judge*.

RASH (M.). Synonymous with *rait*.

RASHINGS (S. W.). Loose *dirt* or shaley beds of rock.

RATCHES (L.). *Lifts* (3) of 5 yards in length along a working *face*.

RATTLE (Lei.). To work (*drive* into or *sink* through) with great vigour and energy.

RATTLE-JACK (M.). Carbonaceous shale; also *Hoo Cannel*.

RATTLER (C.).

RATTLERS (Y.). Cannel coal.

REARER (N. S.). See *Edge Coals*.

RECEIVING RODS. Auxiliary *cage guides* at *insets* and at pit tops.

RECK (L.). Chips of wood and other débris.

RED MEASURES. Generally refers to the strata of Permian or Triassic age.

REDD. 1. (S.) To *scour* through, take down, or to *rip*.

2. To clear out pillars of coal.

3. Pit rubbish or débris.

REDD BING (S.). A *spoil heap* on the surface.

REDDSMAN (S.). One who *redds* (1), or works at night in cleaning up and repairing roadways, &c.

REED (S.). See *Cleat* (1).

REFUGE HOLE. A place formed in the side of an underground plane or horse road, about three feet square and five or six feet high, in which men can take refuge during the passing of a *train*, or when firing *shots*. They may not be put in more than 20 yards apart on engine planes, or 50 yards where horses are employed.

REGULATOR. A *door* in the mine, the opening or closing of which regulates the supply of *ventilation* to a *district*.

RELEVÉE (Pr.). A certain thickness of coal beds and intervening *measures* (varying between 88 and 160 yards) in inclined strata, which forms a *lift* (10) or series of *workings* being prosecuted to the *rise* at one time. They are carried on on both sides of the *shafts* and there are generally three in course of being worked one above another simultaneously, viz. the uppermost which is nearly *worked out*, the middle one in full

Fig. 109.

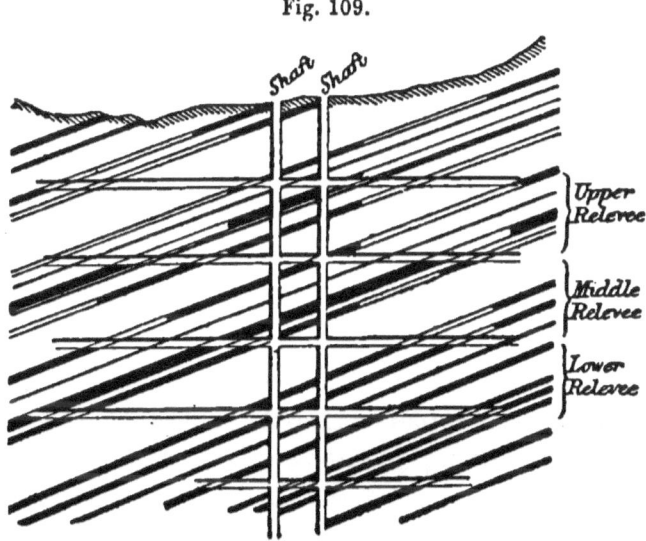

swing, and a lower one in course of being formed to take the place of the upper one. See Fig. 109.

REMBLAIS. 1. (F.) A system of working a very thick (sometimes 80 ft.) seam in Central France. A horizontal slice is first taken out 6 feet 6 inches in height across the seam, and the space filled up with stone, &c., brought from the surface. A second *lift* (3) is then extracted, and so on.

2. (F.) Synonymous with *long-wall*.

3. (F.) Synonymous with *goaf*.

RENK or RANK (N.). A standard measurement of length employed underground, being 60 to 80 yards, measured off periodically by an *overman*.

RENT (S.). See *Back* (1).

REPAIRER. A man who works in the mine, generally at night, setting *timbers*, *pack*-building, *road* (2) laying, &c.

RETURN. The *air-course* along which the vitiated air of the mine is returned or conducted back to the *upcast* shaft.

RETURN AIR. The *air* or *ventilation* which has been passed through the *workings*.

REVERSE FAULT. See *Overlap*. See *Fault*, Fig. 70 (5).

REVIERBEAMT (Pr.). The chief *Inspector* of a district who gives actual decisions, subject to appeal, in reference to mining questions, rules, &c. He receives every year from the *coal master* a *plan* of the *workings* proposed to be carried out during the following year, to which he may object within 15 days. He acts under the authority of the *Oberbergamt*.

RHONE (S.). A *trow* or gutter, generally 12 feet in length.

RIB. 1. A narrow strip or block of solid coal.

2. (S.) A *seam* or stratum.

RIB AND PILLAR (S. S). A system upon which the Thick coal seam was formerly extensively mined, being a kind of *pillar and stall* plan.

RIBAND-STONE. Sandstone in thin layers alternating in colour, generally light and dark grey.

RIBBING. 1. (L.) A strip of coal three yards in width.

2. Enlarging a *heading* or *drift*.

RIBS (Pa.). The sides of a rectangular *pit-shaft*.

RICE (B.). See *Lacing* and *Lagging*.

RICING (N. S.). See *Lacing*.

RICKET or RICKETING. 1. (M.) A narrow *brattice* for ventilation. See Fig. 27 (right-hand side of).

2. (M.) A channel formed along the *floor* of a mine for drainage purposes.

RIDDING. 1. (N.) See *Redd*.

2. (N.) Separating ironstone from coal shale.

RIDDING PUCKING (S. W.). Cutting up a *crept floor*.

RIDE. To be in a *cage* or *bowk* whilst descending or ascending a *pit-shaft*, or to ride in *trams* on *planes* or *ways*.

RIDER. 1. A guide-frame for steading a *bowk* in a *sinking pit*.

2. (S. W.) Lads who ride upon the *trams* on *engine planes*.

3. A name commonly given to a thin seam of coal overlying a thicker one.

RIFLING (S. S.). Working the upper portion of a coal seam over a *waste* or *goaf*.

RING. 1. A complete circle of *tubbing plates* placed round a *pit-shaft*.

2. (N.) See *Garland* (1).

RING-CRIB. A *wedging crib* upon which *tubbing* is placed, having a gutter or ring cast round the inner edge, to collect any water that may run down the walls of the *shaft*.

RINGER (D.). A hammer for driving wedges.

RINGER AND CHAIN (M.). See *Dog and Chain*.

RINGES (N.). See *Cowls*.

RIP (M.). To cut or blast down the *roof* or *top*.

RIPPER. A man who *rips*.

RISE. 1. The inclination of strata when viewed in the direction opposite to the *dip*.

2. An increase of wages paid to colliers, &c.

RISER (N.). An *upthrow fault*.

RISE SPLIT. A proportion of the ventilative current sent into a *rise district* of a mine.

RISE WORKINGS. Underground *workings* carried on to the *rise* or high side of the *shafts*.

RISING MAIN. See *Column* in re water.

RIVELAINE (Belg.). A *pick* much used by *colliers* (1).

RIVES IN. Cracks open, or produces fissures.

ROAD. 1. Any underground passage, way, or gallery. See *Main Road*.

2. The iron rails, &c., or Permanent Way of underground *roads* (1).

ROAD-HEAD (S.). See *Gate-end*.

ROADING. Repairing and maintaining *roads*.

ROB. To cut away or reduce the size of *pillars* of coal, &c.

ROBBED OUT (C.). Worked away. See *Hollows*.

ROBBLE. A *fault*. See *Horses*.

ROCK. Generally means sandstone.

ROCK AND RIG (S. S.). A sandstone full of little patches and shreds of coal, sometimes mixed up in a very singular way.

ROCK BIND or ROCK BINDERS. Sandy shale.

ROCK DRILL. A rock-boring machine worked by hand or by compressed air or by steam. Very extensively employed in *tunnelling, sinking*, and *driving stone-drifts* in mines.

ROCK FAULT. A replacement of a *coal seam* over a greater or less area, by some other rock, usually sandstone. They may be regarded as ancient stream courses. Are narrow as compared with their length, and turn and wind about as do rivers. See Fig. 70 (2), which is a *rock fault* in cross section.

ROCK HEAD (Ch.). The uppermost stratum of the rock-salt beds.

ROCKING LEVER. See *Brakestaff*.

RODDING. The operation of fixing or repairing wooden *cage guides* in *shafts*.

RODS. 1. Vertical or inclined timbers for actuating pumps.

2. Long iron bars of Swedish iron of the toughest quality, for boring through rocks, &c.

3. See *Cage Guides*.

ROLL. 1. An inequality in the *roof* or *floor* of a mine.

2. (S. W.) The *drum* of a *winding engine*.

3. See *Bump*.

ROLLER. Small steel, iron, or wood wheel, upon which a *hauling* rope is carried just above the *floor*. They are placed every 8 or 10 yards along an *engine plane*. They are from 4 inches to 12 inches in diameter, and in length or width from 1 inch to 24 inches.

ROLLEY (N.). A kind of truck running upon wheels for carrying *tubs* or *corves*, drawn by horses along underground *ways*.

ROLLEY-WAY (N.). The underground road along which *rolleys* are conveyed.

ROOF. The top of any subterraneous passage or working.

ROOFING (Ch.). The upper 5 or 6 feet of the rock-salt beds.

ROOM. 1. (S.) A *heading* or short *stall*.

2. A weight of 7 tons of coal, or $5\frac{1}{4}$ chaldrons by measure.

ROOM AND RANCE (S.). A system of working coal somewhat similar to *double stall*, which see.

ROOVE. To rub or knock against the roof.

ROPE-ROLL. The *drum* of a *winding engine*.

ROSH (Lei.). See *Rait*.

ROTCHE or ROCHE (S. S.). A softish and moderately friable sandstone.

USED IN COAL MINING, ETC. 207

ROUND COAL. Coal in large lumps, either hand-picked or after passing over *screens* to take out the *small*.

ROW (N. S.). A *seam* or *bed* (2), e. g. the "Rowhurst" and "Two Row" coals.

ROYALTY. 1. The mineral estate or area of a colliery, or a portion of such property. A field of mining operations.

2. A rent payable on coal, &c., worked from a *Royalty* (1). See *Acreage Rent*.

RUBBING SURFACE. An expression used in reference to *ventilation*, meaning the total area of a given length of *airway*, i. e. areas of sides, top, and bottom, all added together.

RUBBISH. Fallen stone from the *roof*, *holing dirt* and débris made in *sinking*, *dinting*, &c.

RUBBLE. A coarse gravelly loose stone or bed of rock.

RUBBLES. 1. (F. D.) See *Kibbles* and *Nuts*.
2. (S. W.) Slack or *small*.

RUCK (L.). The stock of coals on the *bank* (1).

RUDDING (N.). See *Redd*.

RUN. 1. See *Journey*.
2. To *brake* or *jig*.
3. A breakaway upon an *inclined-plane*.
4. (Pa.) The sliding and crushing of pillars of coal, producing *falls* of *roof*.
5. A word commonly made use of to express the degree of leverage or breaking-down power of a *shot*. When a considerable length of *wall face* is brought down

by the action of a single shot, the shot is said to *run* well.

6. To work a *winding* or *hauling* engine.

7. Soft ground is said to *run* when it becomes mud and will not hold together or stand.

RUN COAL. Soft bituminous coal.

RUNNER. 1. A movable bridge or platform over the mouth of a *sinking pit*.

2. A *fault slip*.

3. A *Crow's-foot*, which see.

4. (Y.). A flat piece of timber placed above *bars*, and connecting them.

5. (Lei.) The piece of timber placed in a horizontal position between the two inclined sprags in *cockermegs*. See Fig. 42. It is cut from two to four feet in length, and assists greatly in steadying the *sprags* and to keep up the coal *wall*.

RUNNER ON. See *Bottomer*.

RUNNING AMAIN (S.). The breaking and running of a *winding rope* down into the *pit-shaft*.

RUNNING A MINE (S.). Forming a *drift* (2).

RUNNING GUG (Som.). A self-acting *jig*.

RUNNING LIFT. A *sinking sett* (1) of pumps constructed to lengthen or shorten at will, by means of a sliding or telescopic *windbore*.

RUNNING MEASURES. Sands and gravels containing much water.

RUNNING THE DRUM. The lowering or sinking of a cylinder or drum through *quick* ground, to secure the upper part of a coal *shaft*.

Run Rider. A lad who goes with a *train* on an *engine plane*.

Run the Tow (S.). Sliding down the *pit-shaft* on the *winding rope*. Running the tow is a common practice in shallow mines.

Rush (S.). The sudden *weighting* of the *roof* when *robbing the pillars* begins, and the *roof* is a strong one.

Rusks (N.). Small slack, or that next larger than *dust* or *dead small*.

Ruttles (Y.). Shattered and faulty *ground* running roughly parallel to the plane of a *fault*.

## S.

Saddleback. A depression or valley in strata. See *Roll*.

Safety Cage. A *cage* fitted with an apparatus for arresting its motion in the *shaft* in case the *winding rope* breaks.

Safety Door. A strongly-constructed *door* hinged to the *roof* of the mine, and always kept open and hung near to a *main door*, for immediate use in case of damage by *explosion* or otherwise to the *main door*.

Safety Lamp. A miner's lamp which reveals the presence of *fire-damp* when the proportion of this gas in the atmosphere of the mine is such that the mixture is already very dangerous, and the moment of explosion is near at hand. The flame is generally surrounded by a cylindrical covering of wire gauze, which protects the surrounding atmosphere from being fired, even though the gases within the lamp have reached the explosive proportions. See *Clanny, Davy, Geordie, Mueseler* (Fig. 110).

Fig. 110.

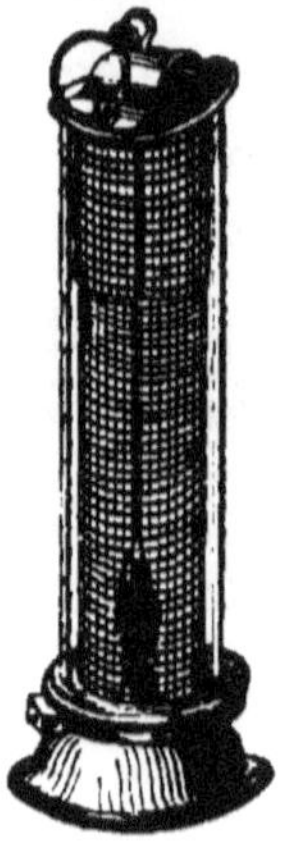

SAFETY TOOLS. Consist of Catching Hooks, Grappling Tongs, *Fish-heads, Bell-screws*, and the like, for recovering broken *boring* tools, picking up material, &c., at the bottom of *boreholes* (1) and *Kind-Chaudron sinking pits*.

SAGGER or SEGGER. A kind of *fireclay*.

SALTING. Sprinkling salt upon the *floors* of underground *ways* in very dry mines, in order to lay the dust. See *Coal Dust*.

SAMPSON POST (Pa.). A stout wooden post carrying the working beam of a *boring* apparatus.

SAW. A tool for removing irregularities from the sides of *boreholes* (1).

SAWNEY (M.). To lower full *trams* down a *road* or *face* that *dips*, with a rope or a chain for a brake, or drag, passing round a *prop*, &c.

SCALE. A small portion of the ventilative current in a mine passing through a certain-sized aperture.

SCALE DOOR. See *Regulator*.

SCALLOP. To cut or break off the sides of a *heading* without *holing* them, or using powder.

SCAMMED (N.). Sooty.

SCAMY-POST (N.). Soft, short, jointy freestone, thinly laminated and much mixed with mica.

SCARES (N.). Thin laminæ of iron pyrites or spar in coal.

SCATTER (Y.). A rumbling or falling noise in a *pit-shaft*.

SCISSORS FAULT. A fault of dislocation in which the beds are thrown somewhat as shown in Fig. 111.

Fig. 111.

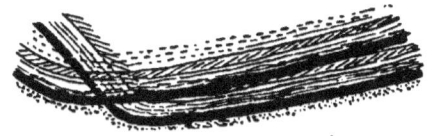

SCOOP (Y.). A barrel or box used in a *gin pit*.

SCORE. 1. (N.) A standard number of *tubs* of coal upon which *hewers'* and *putters'* prices for working are paid. The *score* generally varies between 20 to 26 *tubs*.

2. A bill run up by a *collier* (1) in " bad times " for the necessaries of life.

SCOTCH. 1. A wooden stop-block or iron catch placed across or between the rails of underground roadways, to keep the *tubs* from running loose, or to hold them when standing upon an *inclined plane*.

2. (Lei.) The lower *lift* (3) of coal which is wedged up in *driving* a *heading* a few yards from the *back* (2). By having a *scotch* formed, it enables four *hewers* to work together in *driving* a *heading*, say 7 feet by 6 feet.

SCOTCH GAUZE LAMP. See *Gauze Lamp*.

Scour (M.). To excavate or *brush* a roadway through a *goaf*.

Scovens (S. S.). Forks (?) for filling coal into *tubs*.

Scowl a Brow (F. D.). To *drive* a *heading* or level by guesswork.

Scowles or Scoules (F. D.). See *Meend*.

Scraper. A light wooden rod for clearing *boremeal* out of a drill hole.

Scratcher. A boring tool for scraping or scratching up the débris, to be afterwards removed by a *mizer*.

Screen. 1. A mechanical apparatus (a sort of gridiron) for separating small from large coals. It is erected on the surface.

2. A cloth *brattice* or curtain hung across a road in a mine to direct the *ventilation*.

Scrin (D.). *Ironstone* in irregular-shaped nodules.

Scroll Drum. See *Conical Drum*.

Scronge (S. W.). The loosened or broken strata overlying and produced by *workings* underneath.

Scud. 1. (Lei.) Very thin layers of soft matter, such as clay, sooty coal, &c.

2. (M.) Iron pyrites embedded in coal seams.

Sea Coal. That which is conveyed away from the collieries by sea; be it *house, steam*, or manufacturing coal.

Sealing. Shutting off a pit or part of a mine after a fire or an *explosion* by means of *stoppings*.

Seam. Synonymous with *bed, mine, vein, row, band,*

&c. Some *seams* are made up of a number of *beds* interstratified with shale, &c.

SEAT (Y.). The bottom or *floor* of a mine.

SEAT EARTH (Y.). Generally a kind of hard *fire-clay* forming the *floor*.

SEATING. The masonry in which a steam boiler is set.

SEAT STONE. See *Seat Earth*.

SECOND WORKING. The operation of getting or working out the *pillars* of coal formed by the *first working*; e. g. *long-wall*, *working home*, working the *broken*, *drifting back*, &c. *Second working* is paid for by the ton or by the *score* (1).

SECTION. 1. A term usually applied to a vertical exposure of strata.

2. A drawing or diagram of the strata sunk through in a *pit-shaft* or inclined plane, or proved by *boring*.

SEED BAG (Pa.). A stout leather tube passed with the tubing or lining of a *borehole* (1) into water-bearing ground. The annular space between the tube and the leather is filled with flax seed, which, becoming moist with the water, expands, and thus effectually stops out the water.

SEG (N.). To bend down in the middle.

SELF-ACTING INCLINED-PLANE. An inclined-plane upon which the weight or force of gravity acting upon the full *tubs* is sufficient to overcome the resistance of the *empties*; in other words, the full set (1) draws the empty set up the hill. See *Incline*.

SELF-DETACHING HOOK. See *Detaching Hook*.

SEPARATION COAL. Coals of various sizes loaded separately into wagons, &c. See *Dry Separation* and *Wet Separation*.

SEPARATION DOORS. *Doors* fixed underground (generally two, sometimes three), between the *intake* and the *return*, near the *pit bottom*.

SEPARATION VALVE. A massive cast-iron plate suspended from the *roof* of a *return air way*, through which all the *return air* of a separate *district* flows, allowing the air to always flow past or underneath it; but in the event of an *explosion* of *gas* the force of the *blast* closes it against its frame or seating, and prevents a communication with other *districts*. The *blast* being over, the weight of the valve causes it to return to its normal position, aud allows the *district* to breathe again.

SERVE (N.). *Gas* is said to *serve* when it issues more or less regularly from a *fault slip*, a *break* (1), &c.

SET. 1. (N.) See *Journey*.

2. (S. S.) To get the *sides* off and trim up a *heading*.

3. (N.) To load a *tub* unfairly by placing the greater part of the coals on the top of it and leaving the bottom part comparatively empty.

4. (N.) The natural giving way of the *roof* for want of support.

5. To fix in place a *prop* or *sprag*.

6. Timbers fixed in a *heading*, &c., as in *Double Timber*, which see.

7. To set or make an agreement with miners to do certain work by the bargain: e.g. to set a *stall*.

SET COAL (Lei.). Coal near to *hollows* having a hard dead nature.

SET OUT (N.). See *Lay Out*.

SETT. 1. A *column* of *pump trees*, with *buckets* or ram, &c., complete.

2. The area of mines *worked* (4) by a separate colliery or firm.

3. (M.) A measure of length along the *face* of a *stall*, usually from say 6 to 10 feet, by which *holers* and *drivers* work and are paid. A certain number of *setts* comprise a day's work.

4. Setting up a *dial* for taking a bearing or *sight* (2).

SETTERS (N.). Large lumps of coal placed round the sides of coal dealers' carts, for the purpose of piling up a good load in the centre.

SETTINGS (S. S.). Timbers set as shown in Fig. 59. See *Double Timber*.

SETTLE BOARDS. 1. (N.) Iron plates or sheets forming the floor of a *heapstead*, to admit of the *tubs* being pushed and turned about with facility.

2. (N.) See *Cage Shuts*.

SETTS OFF. See *Distance Blocks*.

SHAB (Som.). Friable shaley rock.

SHAFT. 1. A vertical pit or hole made through strata through which the produce of the mine is brought to the surface, and through which the *ventilation* is passed into and out of the *workings*. It is generally the only outlet from the mine to the surface. Shafts are usually constructed in a circular form, though oval and rectangular ones are not uncommon. They vary in diameter from say 7 to 20 feet. The deepest shaft in Great Britain is 2817 feet, and 16 feet in diameter within.

2. A wooden handle of a *pick*, &c.

3. (S. W.) To pull or draw at a *tub*.

SHAFT FOOT (S.). See *Pit Bottom*.

SHAFT KIP. See *Kip*.

SHAFT LAMP. See *Comet*.

SHAFT PILLAR. Solid coal left *unworked* beneath colliery buildings and around the *shafts*, to support them against subsidence and *creep*. The size and form of *shaft pillars* are regulated by the depth to, and thickness and inclination of, the seam of coal to be worked.

SHAFT RENT. 1. Rent paid for the use of a *shaft* (1) for raising the minerals from another *royalty* by *outstroke*.

2. Interest on capital invested in *sinking* a *shaft* (1).

SHAFT-TUNNEL (N. S.). *Crutts* or levels driven across the *measures* from *shafts* (1) to intersect *rearers*.

SHAGGY METAL (Ch.). See *Horse Beans*.

SHALE. Strictly speaking, all argillaceous strata that split up or peel off in thin laminæ. In mining language it is generally indurated clay or *bind* (1).

SHAM DOOR. A check or *regulator door*.

SHANK (S.). A shallow *shaft* (1) underground.

SHARP (M.). Hard and compact in re rock or sandstone.

SHARP GAS. *Fire-damp* which explodes suddenly within a *safety lamp* without showing any perceptible *cap* (1).

SHEARER. See *Saw.*

SHEARING. Cutting a vertical groove in coal similar to *holing* at the bottom of the seam.

SHEAR LEGS. A high wooden frame placed over an *engine* or pumping *shaft* (1), fitted with small pulleys and rope for lifting heavy weights in the pit.

SHEARS (S.). A *haulage clip*, which see.

SHED. 1. (Pa.) A kind of long car or trolley.

2. A thin smooth *parting* in rocks, having both sides polished.

3. A very thin layer of coal.

SHEETS. Coarse cloth curtains or *screens* (2) for directing the ventilative current underground.

SHELL BAND. See *Mussel Band.*

SHELL DOOR. A temporary *door.*

SHETH. 1. (N.) To *course* the air in the *workings.*

2. A set or panel of *boards* (1).

SHETH DOOR (N.). A *door* fixed in a *working* going *headway course*, for temporary purposes only.

SHETHING THE AIR (N.). Ventilating the *goaves* in a systematic way.

SHETHS (N.). The ribs of a chaldron wagon.

SHE'S FIRED! An *explosion* of *fire-damp* has taken place in the pit! See *Squat Lads!*

SHEUGH or COAL-SHEUGH (S.). A *shaft* (1) or coal pit.

SHIDES (B. S.). Pumps for draining mines.

SHIFT. 1. A certain number of hours of work; a

certain proportion or change of workmen. See *Double Shift*.

2. A *fault* of dislocation.

SHIFTER. 1. See *Runner on*.

2. (N.) One who *repairs* roadways in a mine.

SHIFTWORK. Work performed underground: e.g. *timbering, way* (1) cleaning, &c.

SHIVER. See *Bind* (1).

SHIVERED. Knocked to *small* by blasting.

SHIVERY. Short and tender; easily broken up or *worked* (5).

SHOE-NOSE SHELL. A *cleanser* specially constructed for working in hard ground.

SHOES. Steel or iron guides fixed to the ends and sides of *cages*, to fit and run upon the *conductors*.

SHOE SHELL. A tool used in deep *boring* for cleansing out the *boremeal*. It has a valve at the bottom, opening upwards.

SHOOTING. Blasting in a mine.

SHOOTING FAST (L.). Blasting without previously *holing* or *shearing* the coal.

SHOOTING THE GOB (N. S.). Working the coal in the pillars of *rearers* by blasting.

SHORN. Cut with a *pick*.

SHORT (N. S.). Coal is *short* when of a very friable or tender nature.

SHORTS. 1. The contents of *trams* filled with coal, or coal and *dirt* mixed, otherwise than in accordance with the colliery regulations.

USED IN COAL MINING, ETC. 219

2. Deficiency of mineral worked under a lease during any year or other period agreed upon. In granting a lease of coal, &c., it is customary to insert a clause which provides that if the quantity of coal raised from the estate during any year at an *acreage* or *tentale* rent does not amount to the certain or *minimum* rent, the lessee may in any subsequent year get and raise such quantity of minerals as shall make up the deficiency without paying any more rent than the *minimum*. Exercising this right is commonly known as *making up shorts*.

SHORT STALL (M.). See *Single-road Stall* (Fig. 113).

SHORT-WORKINGS. See *Shorts* (2).

SHOT. The firing off of a cartridge of gunpowder, dynamite, &c., in blasting.

SHOT FAST. Coal which is worked by blasting, and has had a *fast shot* in it.

SHOT HOLE. The *borehole* (2) in which the explosive substance is placed for blasting. It is usually from 18 inches to 3 feet in depth, according to the nature of the rock (including coal) being operated upon and from 1 inch to 1¾ inches in diameter. These holes are put in either by hand or by machinery. There are hand-power rock perforating machines, both percussive and rotary in action, also similarly acting machines worked by steam and compressed air. Hand-made holes with the ordinary drill or *jumper* are always more or less three-cornered in shape.

SHOT LIGHTER or SHOT FIRER. A man specially appointed by the *manager* of a mine to fire off every *shot* in a certain number of *stalls* or *heads* during the *shift*. He shall not *fire* until he has examined the immediate neighbourhood of the shot and found it free from *gas* and otherwise safe.

SHOULDER CUTTING (S. S.). Cutting the sides of the upper *lift* of a *working place* in a Thick-coal colliery next the *rib*, preparatory to *falling* the coal.

SHOW. When the flame of a *safety lamp* becomes elongated or unsteady, owing to the presence of *fire-damp* in the air, it is said to *show*.

SHUT or SHUTT. 1. (S. S.) The crushed and broken-down roof or overlying rock of a seam of coal.

2. Old workings. See *Goaf*.

SHUTERS (S. S.). *Blue Bind*.

SHUTTER. 1. A movable sliding door having balance weights attached, fitted within the outer casing of the Guibal *fan*, for regulating the size of the opening from the fan, to suit the ventilation and economical working of the machine.

2. The vibrating arm or door of the Cooke *Ventilator*. See *Ventilator*.

SHUTS (S.). See *Keps*.

SHUTTLES (L.). Natural cracks running at right angles to the *dip* of the strata.

SHUTTING. See *Shooting*.

SIDDLE (N.). The inclination or *dip* of a bed of coal, &c.

SIDE. 1. The more or less vertical face or wall of

coal or *goaf* forming one side of an underground *working place*.

2. (L.) A *district*.

SIDE CHAIN (M.). A chain hooked on to the sides of *tubs* when running upon an *engine-plane* or *jig*, to keep all the tubs together in case a coupling breaks.

SIDE OF WORK (S. S.). A kind of chamber or panel in the Thick-coal *workings* containing from two to twenty *pillars*. Fig. 112 shows a plan of a *side of work*.

Fig. 112.

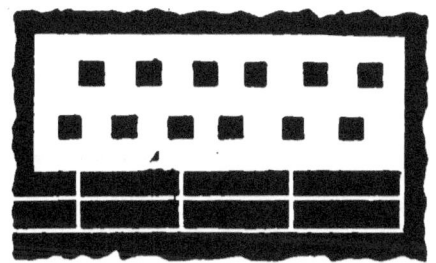

SIDE-OVER (N.). To cut or *drive* in a line with the *cleat* through a pillar of coal when working the *broken*.

SIDE-WAFER or SIDE-WAVER (N.). 1. Overhanging stones or roof in underground roads liable to drop.

2. A fall of *sagger*, &c.

SIDING-UP (N.). Width of a *tub* and room for *gears* (1).

SIGHT (Eye-sight). 1. On reaching a pit bottom, the eyes require to be allowed time to adjust themselves to the darkness. This period is known as taking time to *get your sight*.

2. A bearing or angle taken with a *dial* when making an underground survey.

SIGNS (employed upon colliery working plans):—

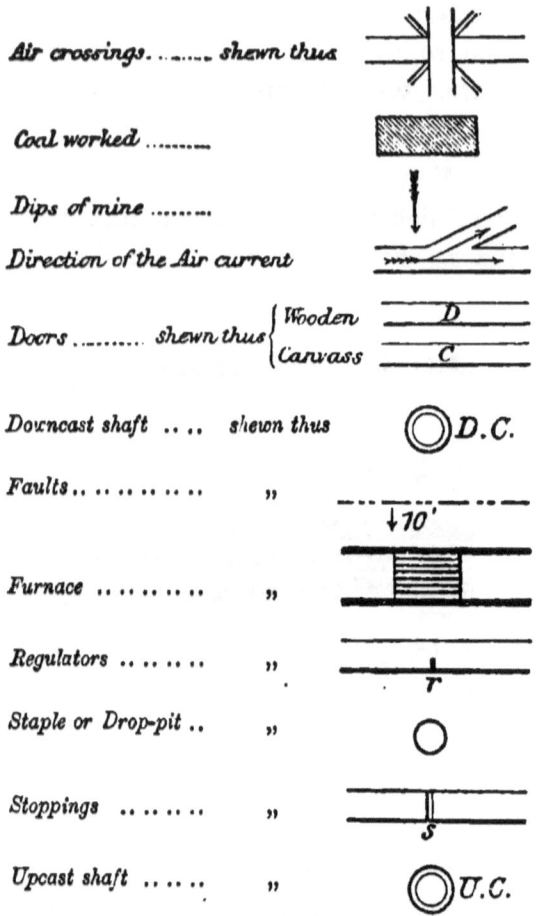

SILL. 1. (N.) A face of hard rock: e. g. the Great Whin Sill.

2. (C. Y.) Much the same as *Clunch, Spavin, Warrant,* &c.

Sing. When a freshly cut-into seam of coal gives off *gas* and water with a hissing noise resembling the boiling of a tea-kettle, it is said to *sing*.

Singing Coal. A bed of coal from which *gas* is ordinarily issuing from the partly-exposed *face* in the mine, producing a hissing sound, particularly if the surface be wet. This is the usual manner in which *gas* is given off in mines.

Singing Lamp. A *safety lamp* which, when placed in an atmosphere of explosive *gas*, gives out a peculiar sound or note, the strength of the note varying in proportion to the percentage of *fire-damp* present.

Single-road Stall (S. W.). A system of working coal as shown in plan, Fig. 113.

Fig. 113.

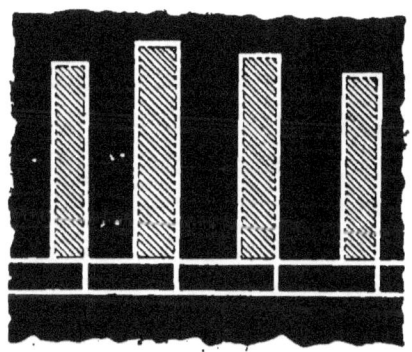

Single-rope Haulage. That system of underground *haulage* in which a single rope is used, the empty *set* (1) running *inbye* by gravity.

Sink. 1. To excavate strata downwards in a vertical line, for the purpose of *winning* and *working* minerals.

2. To *bore* (1) or put down a *borehole* (1).

SINKER. A man who works at the bottom of a *shaft* in course of being *sunk*. He bores the *shot-holes*, charges them and fires them off, sends the débris to *bank*, and assists in putting in *tubbing*, *walling*, pumps, &c.

SINKING. A *pit-shaft* or *shafts* (1) being put down in order to work coal, &c.

SINKING PIT. A *shaft* in course of being sunk. See *Sink*.

SINKS (L.). Natural cavities met with in iron mines.

SIT (M.). A coal *face* (1) or *buttock* is said to *sit* when, after the *sprags* have been drawn, it will not fall over and break up, but merely cracks off and rests in that position until pulled over.

SITS. 1. (S.) *Creeps* or subsidences of *cover*.
2. A fall of *roof*.

SIZE. In reference to a *fault*; this word means the extent of the displacement or the *throw*, which see.

SKEEL (Som.). A kind of *cage* in which coals are lowered down the *cuts* or *staples*.

SKEP. A bucket or tub a pit-horse drinks out of.

SKERRIES (W.). Greenish-white micaceous sandstone.

SKERRYSTONE (M.). Hard, thin-bedded sandstone.

SKEWS (S.). See *Lypes*.

SKID (B.). See *Hudge*.

SKIDS. Slides or slippers upon which certain coal-cutting machines travel along the *faces* (1) whilst at work.

SKIP, sometimes SKEP. 1. (S. S.) A coal *tram* or *box*.

2. See *Cuffat*.

3. (S. W.)

SLABS. *Lagging* placed over *bars*.

SLACK. Small coal which will pass through a *screen* (1). There is no standard size distinguishing *coal* (2) from *slack*.

SLAG (N.). See *Brat*.

SLANT. An underground roadway driven more or less on the *rise* or *dip* of the mine.

SLAP (Som.). See *Slack*.

SLATCH (Som.). See *Lathe*.

SLATE COAL. A hard, dull variety of coal, not unlike *Cannel*.

SLED, properly SLEDGE. See *Cart*.

SLEEK (B.). Soft and troublesome, as applied to the state of the *floor* in *steep seams*.

SLEW (D.). See *Lum*.

SLICKENSIDES. The smooth striated surface of joints on opposite walls of a *fault* or fissure.

SLICKS. Smooth *partings* or mere planes of division in strata.

SLIDE. A *fault*.

SLIDES. See *Cage Guides*. Made either of wood or rolled iron.

SLIDING JOINT. A *boring rod* made in two portions, one sliding within the other, to allow of the concussion or shock produced by the weight of the falling rods being modified or taken off the cutting tool in very deep *boreholes* (1).

SLIDING SCALE. A mode of regulating the amount of wages in mining districts by taking as a basis for calculation the market value of coal or iron, the amount rising and falling with the state of the trade. For example, when pig-iron sells for (say) 60s. per ton, the wages of underground men to be (say) 5s. a day; but when pigs are at 70s., miners' wages shall be (say) 5s. 6d. a day, or rising 6d. a day for each rise of 10s. in the price of iron.

SLIDING WINDBORE. The bottom pipe or suction-piece of a *sinking sett* of pumps (pumps used in a *sinking pit*), having a lining made to slide or telescope within it, to give length without altering the adjustment of the whole column of pumps.

SLIG or SLIGGEN (I.). Shale.

SLINE or SLYNE. 1. A facing or smooth *parting* or joint in coal, &c.

2. (M.) *Potholes* in the *roof*.

SLIP. 1. A fault. See Fig. 70 (1).

2. A smooth joint or crack in strata.

SLIP CLEAVAGE (S. W.). The *cleat* of the coal running in planes parallel with *slips* (1). See Fig. 114.

Fig. 114.

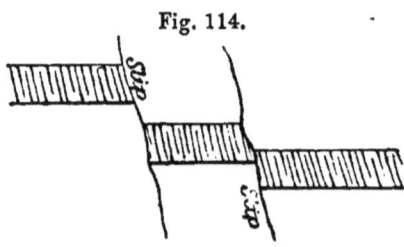

SLIP-DYKE (N.). See *Slip* and *Fault*.

SLIPE (S. S.). A *skip* without wheels, a sledge.

SLIPPERS. See *Shoes*.

SLIPPY (M.). Full of *slips* (2).

SLIPPY BACKS (N.). Vertical planes of cleavage occurring every four or five inches in the seam of coal.

SLIP SPEAR (Pa.). A tool for extracting tubing from a *borehole* (1).

SLIP-THINGS (S. S.). The more or less vertical planes of cleavage in coal, &c.

SLIP-TROUBLE (S.). See *Slip* (1).

SLIT. A short *heading* put through to connect two other *headings*.

SLITTER. See *Pick*.

SLIVERS. Strips of wood or iron fitted in between the edges of boards in wooden *bratticing*, to make the joints air-tight.

SLOOM (M.). A softish earthy clay or shale often underlying a bed of coal.

SLOPE. 1. See *Slant*.

2. (Pa.) The main *engine-plane* or inclined roadway driven in the seam of coal worked from the surface *outcrop*, up which the whole of the produce of the mine is raised by the *winding engine*.

SLOT (Y.). To *hole* (1).

SLOTTINGS (Y.). Coal cut away in the process of *holing*.

SLUDGE PUMP. A short iron pipe or tube fitted with a valve at the lower end, with which the *boremeal* is extracted from a *borehole* (1).

SLUDGER. See *Sludge Pump*.

SLUM, SLUMS, SLUMBS. 1. (N. S.) A blackish, slippery, indurated clay.
2. A soft clayey or shaley bed of coal.

SLYPE (S.). See *Sawney*.

SMALL. See *Slack*.

SMART FIRE (N.). A severe though small *explosion*.

SMART MONEY (N.). A weekly allowance of money given by employers to workmen who get injured whilst at work.

SMELL. The early indication of a *fire-stink* perceptible to the nose.

SMIFT. A bit of touch-paper, touch-wood, greased candlewick, or paper or cotton dipped in molten sulphur, attached by a bit of clay or grease to the outside end of the train of gunpowder when blasting. Its object is to ignite the *shot* after giving the miner sufficient time to retire to a place of safety.

SMITHEM or SMYTHAM. 1. (M.) Fine slack.
2. Clay or shale between two beds of coal.

SMITH ORE (F. D.). See *Brush* (2).

SMOKY PIT (M.). An *upcast* shaft with a *furnace* at the bottom of it.

SMOOTH (S.W.). The line of *face* (1) of a *stall*.

SMOOTH-HEADS (Y.). See *Bright-heads*.

SMOOTHS (S. W.). Planes of cleavage more or less vertical.

SMUDGE. See *Smithem* (1).

SMUT. See *Coal Smut*.

SMUTH or MUCKS. Very inferior coal.

SNAP (M.). See *Bait*.

SNAPPING TIME (M.). A short period of rest during a *shift* in which a collier takes his *snap*.

SNAPS (M.). A *haulage clip*. See Fig. 79 for tail rope clip.

SNECK Y. A *carving* (2)?

SNECKS (S.). Appliances for diverting wagons from the main line into a siding.

SNIBBLE (N.). See *Locker*.

SNOREHOLES. The holes at the bottom of a *snorepiece* through which the water enters to the pump.

SNOREPIECE. The lowest end of a pump *sett* (1) through which the water passes.

SNUFF. See *Smift*.

SOAPSTONE (Y., N.W.). A variety of *fireclay*, sometimes applied to *Bind* (1).

SOAMS (N.). A pair of cords about three feet in length, by which *foals* and *half marrows* pull *tubs* along the roads.

SOCKET. The innermost end of a *shot hole* not blown away after firing.

SOCKET BAR. See *Beche*.

SODS (Lei.). Clay beneath coal seams.

SOFT. Tender, full of slips and joints, friable.

SOFTS (M.). Coals which easily break up.

SOLE. A piece of timber set underneath a *prop*.

SORTING (M.). Turning over by hand and examining the *round coal* as it comes from the mine; dividing it up according to size and quality into various sorts to suit the trade, carefully throwing aside all inferior or stony coal.

Sos (S. S.). To sink into the *floor* under great pressure from overlying strata.

SOUFFLARD (F.). See *Blower*.

SOUNDING. Knocking on the *roof*, &c., to ascertain if it is sound or safe to work under.

SOUTÈNEMENT (F.). Propping and packing the *roof*.

SPAN-BEAM. A long wooden beam supporting the head pivot of the drum axle of a *gin*, and resting at the extremities upon inclined legs.

SPARE (N.). A deal wedge from 6 to 8 inches long, for driving behind *tubbing plates* when adjusting them to the circle of the *shaft*.

SPAVIN (Y.). Clunch, or ordinary bottom or *underclay*.

SPEAKING-FLAME LAMP. See *Singing Lamp*.

SPEAR PLATES. Wrought-iron plates bolted to the sides of *spears* where joined together. See Fig. 115.

Fig. 115.

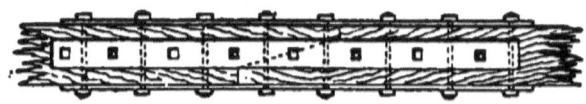

SPEARS. Wooden pump-rods of Memel or pitch pine timber cut into lengths of about 40 feet, and, for heavy work, often measuring 16 inches square. Wrought-iron *pears* are also used.

SPIDERS (U. S. A.). See *Drum Rings*.

SPIKING CURBS. Light rings of wood to which

planks are spiked, bevelled to suit the sweep of the *shaft*, when *plank tubbing* is used in sinking through water-bearing ground.

SPILES. 1. Narrow-pointed *tubbing* wedges.
2. See *Lacing*.

SPILING (N.). See *Spiles* (1).

SPIRAL DRUM. See *Conical Drum*.

SPIRES (Lei.). Coal of a hard, dull, slaty nature, and difficult to break up.

SPIRAL WORM. A tool for extricating broken *boring rods*. Fig. 116.

Fig. 116.

SPLINT or SPLENT (S.). A laminated, coarse, inferior, dull-looking, hard coal, producing much white ash; intermediate between *cannel* and common *pit coal*.

SPLIT. 1. A division of the air-current underground. Each separate *district* should have its own *split* of fresh air.

2. To divide the ventilative current after it reaches the *pit bottom*.

3. To divide a *pillar* or *post* (1) by *driving* through it one or more roads.

SPLITTINGS (L.). Two horizontal level *headings* driven through a *pillar* in *pillar workings*, in order to work away the coal in the *pillar*.

SPOIL. Débris [stone, shale, bad coal, dirt, and all rubbish] raised from the mine and thrown on one side.

2. A stratum of coal and *dirt* (1) mixed.

SPOIL-BANK or SPOIL-HEAP. The place on the surface where *spoil* (1) is deposited.

SPOUT (S. S.). A short underground passage in the Thick-coal workings connecting a *main road* with an *air-head*.

SPOUT-HOLE (S. W.). 1. A short siding upon which *trams* are loaded in the pit.
2. See *Bolt*.

SPRAG. A short wooden prop set in a slanting position for keeping up the coal during the operation of *holing*. It is a general rule that sprags shall be set not more than 6 feet apart.

SPRING BEAMS. Two stout parallel timber beams built into a Cornish pumping-engine-house, nearly on a level with the engine beam, for catching the beam, &c., and preventing a smash in case of a break down in the *pit work*.

SPRING DART. An arrow or fish-headed boring tool for extricating a lost implement, or for withdrawing lining tubes. Fig. 117.

Fig. 117.

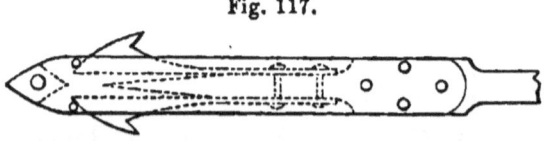

SPRING HOOK. An iron hook attached to the end of a winding, capstan, or crab rope, fitted with a spring for closing the opening, and thus preventing the *kibble*, &c., from falling off.

SPRING POLE. A fir pole having considerable elasticity, to which the *boring rods* are suspended when boring for coal, &c. Also sometimes employed for shallow pumping, when it is actuated by cams or cranks from an engine.

SPUNNEY (L.). See *Jinney.*

SPURNS (S. S.). Narrow pillars or webs of coal between each *holing,* not cut away until the last thing before withdrawing the *sprags.*

SPUR ROAD (S.). A branch *way* leading from a main level.

SPURT (F. D.). A peculiar kind of stone, much disintegrated and mixed with colouring matter.

SQUANDER (Y.). To beat or kill (extinguish) an underground fire.

SQUARE WORK. 1. (S. S.) An old system of working the Thick coal by getting the upper beds first and then the lower ones.

2. A system of working a seam of coal by cutting it up into square blocks or *pillars.* See *Stoop and Room.*

SQUAT, LADS! "Fall flat down on the *floor!*" In the early days of coal mining, before *safety lamps* were much used and ventilating was little understood, setting fire to *gas* was a very common thing; so, whenever an *explosion* took place, the colliers shouted to one another, "*Squat, lads!*" so that by lying close to the *floor* they were often able to escape the fire and *blast* in a great measure, as it passed over them. See *She's fired!*

SQUEEZE. 1. See *Creep*.
2. See *Nip*.

SQUIB. A straw, rush, paper, or quill tube filled with a priming of gunpowder, which is passed through the touch-hole into the *cartridge* or charge in blasting, and ignited by means of a *smift*.

S-ROPE. The *winding* (2) *rope* which passes round the under side of the *drum* (1) from or to the *pulley*; so called because it takes the form of the letter S.

STACK. To build up coals, ironstone, &c., into heaps on the surface for winter or other use

STACKER. 1. One who *stacks* coals, &c.
2. (Lei.) A *butty* out of the pit who looked after the unloading of the *boxes* on the *bank* (on behalf of the coal-getters) in the earlier days of mining.

STACK OUT (M.). To dam off or shut up the entrance to a *goaf* by building a wall of stone or coal in front of it.

STADDLE (M.). The foundation of a *pack* in ironstone workings.

STAGE. 1. A platform upon which *trams* stand.
2. The *pit bank*.
3. A certain length of underground roadway worked by one horse.

STAGE PUMPING. Draining a mine by means of two or more pumps placed at different levels in the *shafts* or *workings* in such wise that each intermediate pump receives its water from the pump next below it, and raises it to the next above; and so on to the surface or *adit*.

STAGE WORKING. A system of working minerals by *open hole* in which the various beds are removed in steps or stages in manner shown in section, Fig. 118.

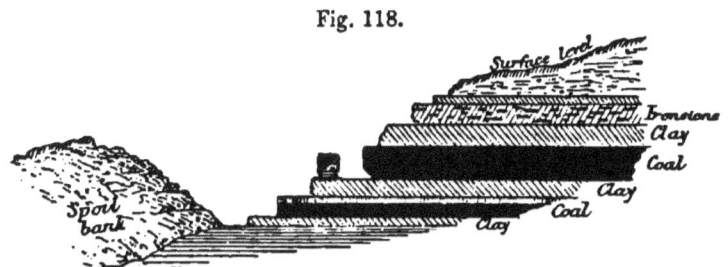

Fig. 118.

STAIR PIT (S.). A shallow shaft or *staple* in a mine fitted with a ladder or steps.

STAITHES (N.). Depôts in which coals are placed when they come from collieries by wagons, to be ready to be loaded into *keels*. They date from 1709. Timber forms the chief material of construction of *staithes*, and they are fitted up with an arrangement of shoots or spouts, down which the coals run into the vessels. See cross-section, Fig. 119. In South Wales hydraulic drops and hydraulic shoots are employed at the *staithes*. When the former are used, the coals, in boxes, are jibbed out, lowered over the vessel's hatchway, and withdrawn again when empty: sometimes a counterbalance weight is employed alone for raising the empty boxes. With the hydraulic shoots, a full wagon is run on to a stage at the top of the shoot, the rear end of the stage is raised or the front end lowered, as the case may be, so as to incline the wagon and cause the coal to fall out at the end door (with which the wagons are all fitted) on to the shoot. Counterbalance shoots also are commonly employed upon *staithes*, wherein all the movements are regulated by counterbalance weights, the action being very similar to that of the hydraulic apparatus above referred to. The coals are sometimes

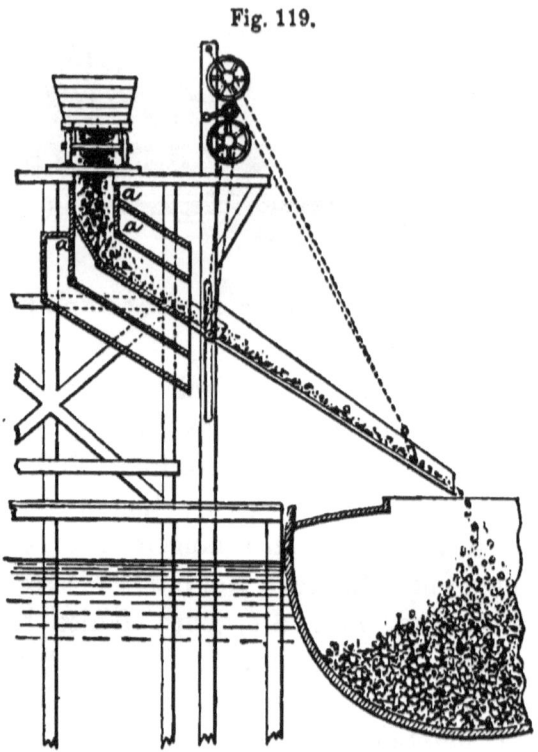

Fig. 119.

lowered from the mouth of the shoots into the bottoms of the vessels by means of an endless band or chain carrying iron buckets, which are fed from a hopper and descend into the hold.

STAKE (Lei.). To fasten back or prop open with a piece of chain or otherwise the valves or *clacks* of a *water barrel* (1), in order that the water may run out of it back into the *sump* when necessary.

STALACTITES (Y.). Icicle-shaped formations upon the *roof*, produced by droppings of water of a saline nature.

STALL. A *working place* in a mine, varying in length from a few feet to 80 yards or more, according to the thickness of the seam and system of working adopted.

STALL AND ROOM WORK. Working the coal in compartments, or in isolated chambers or pillars.

STALL GATE. A *gate road* along which the mineral worked in a *stall* is conveyed to the *main road*.

STALLING. Working in a *stall*, in the capacity of a *butty* or contractor.

STALLMAN. See *Butty*.

STALL WORK. Working coal, &c., in *stalls*.

STAMPING MAUNDRIL (Lei.). A heavy *pick*.

STANCH AIR (Som.). See *Choke-damp*.

STANCHION. See *Puncheon*.

STAND. Does not break down or require timbering. A rock or coal *roof* generally *stands* better than one composed of shale or clay.

STANDAGE. An underground *lodge* or reservoir for water on its way to the *sump* or pumps.

STANDARD. The fixed rate by which colliers' wages are from time to time regulated. See *Sliding Scale*.

STANDARD AIR-COURSES (N.). The various quantities or supplies of fresh air allowed to pass through each *district* or *split*.

STANDING. Not at work, not going forward, idle, at *play* (1, 2), *laid off*.

STANDING BOBBY (N.). An exploded *shot* which rips the coal but does not blow the *stemming* out, and expends itself in *backs* (1) without doing its work.

STANDING FIRE. A fire in a mine continuing to smoulder for a long time; often many years.

STANDING GAS. A body of *fire-damp* known to exist in a mine, though fenced off.

STANDING SET. A fixed *lift* of pumps in a *sinking pit.*

STANK (M.). A water-tight *stopping*; generally a well built brick wall.

STANKING (Ch.). See *Stank.*

STAPLE or STAPLE PIT. A shallow *shaft* within a mine.

STAR REAMER (Pa.). A tool for regulating the diameter of or straightening a *borehole* (1), made star-shaped at the base.

START (N.). A lever for working a *gin* to which the horse is attached.

STATION. 1. Any fixed point underground beyond which *naked lights* may not be carried.

2. Any fixed point in a mine where *deputies* meet to report upon the condition of their respective *districts* and to consult together.

3. An opening into a level heading out of the side of an inclined plane.

STEAM COAL. A hard, free-burning, non-caking, white ash variety of coal. The finest steam coals of South Wales are moderately hard and almost smokeless.

STEAM JET. A system of ventilating a mine by means of a number of jets of steam at high pressure kept constantly blowing off from a series of pipes in the bottom of the *upcast shaft*. Ventilating by this system gives only about 30 per cent. at most of the useful effect produced by a *fan* or *furnace*.

STEEL MILL. An apparatus for obtaining light in the *workings* of a mine where naked lights were considered unsafe. It was brought out by one Spedding, of Whitehaven, in 1760, and used up to 1815, when the *safety lamp* was invented. Its object was to produce a shower of sparks by holding a piece of flint against the rapidly-revolving periphery of a wheel about six inches in diameter, the rim of which was steel. See Fig. 120.

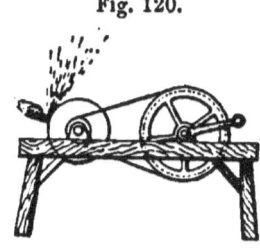

Fig. 120.

STEEP SEAMS. See *Edge Coals* and *Rearers*.

STEER (Lei.). Steep, highly inclined, *dips* fast.

STEIGER (Pr.). See *Fireman*. He has the supervision of only one fixed part or district of a mine.

STEINING. The brick or stone lining of a *pit shaft*,

Fig. 121.

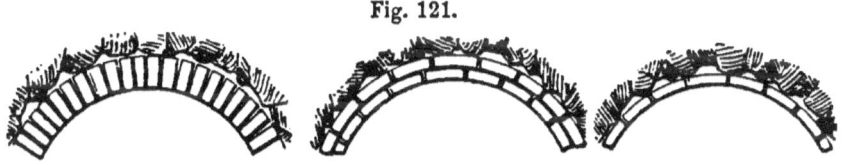

to prevent the loose strata of the sides from falling in. Three methods of *steining* are shown in Fig. 121.

STEMMER. A copper rod used for *stemming* (2).

STEMMING 1. Fine shale or dirt put into a *shot-hole* after the powder, and rammed hard.
2. Ramming or beating the *stemming* (1) solid.

STENTING (N.). See *Stenton*.

STENTON (N.). A short *heading* at right angles to a *cross cut* (2).

STEP BANKS (S. W.). Working places having regular distances along the *carvings* or *cuttings* between the ends of the *stalls* in the *long-wall* system.

STEPPING (N.). The system of working *faces* of coal one in advance of the next to it. See Fig. 91 (upper range of workings).

STEPS. See *Step Banks*.

STERIL COAL. Black shale or clay on top of a coal seam.

STEWARD (Y.). See *Underviewer*.

STIFFENER (S. W.). A *door* for regulating the *ventilation*.

STILLING (N.). The *walling* of a *shaft* within the *tubbing* above the *stone head* (2).

STIMPLES (S. W.). Small timbers. See *Lacing*.

STINT. 1. (M.) A measure of length by which colliers *hole* and cut coal. A stall *face* is usually measured off into a number of *stints* or holing *setts* (3), varying between 4 feet and 6 feet, and each collier *holes* a certain proportion of them for his day's work, according to the length and depth of the *stint*, and hardness of the *seam*.

2. (G.) A certain number of *trams* filled per man per day.

3. (S. S.) A collier's day's work.

4. (B.) To fix upon, or agree to, a certain number of *trams* being filled per *stall* per day.

STIRRUPS. A screw joint suspended from the *brake-staff* or *spring-pole*, by which the *boring rods* are adjusted to the depth of the *borehole* (1).

STOBB. A long steel wedge used in bringing down coal after it has been *holed*. See *Feathers*.

STOCK. 1. Coals laid down at surface during slack trade, or in reserve for an extra demand at any time.

2. The average tonnage sent out of a *working place* in one day.

STOCKING END. 1. (L.) The inner end of a heading at a short distance from which there is a depression or *lum* in the seam, which has become more or less filled with water, causing the *ventilation* to be cut off from the *back* (2).

2. (Lei.) A *Geordie*.

STOMP. 1. (M.) To set a *prop* or *sprag* with one end let into a slight hole cut out of the *floor* or *roof* to receive it.

2. A short wooden plug fixed in the *roof*, to which *lines* are hung, or to serve as a bench-mark for surveys.

STONE. 1. A term commonly used for sandstone, *post* (2), or almost any rock of a stony character.

2. *Ironstone*, which see.

STONE COAL. *Anthracite*, in lumps. Also certain other very hard varieties of coal.

STONEHEAD. 1. A *heading* driven in *stone, bind, measures,* &c.

2. (N.) The first hard stratum met with underlying quicksand.

STONEMAN (N.). "One who is employed in driving a *stonehead*, or who *rips, timbers,* and repairs *roads*.

STONE MINE (S.). An *ironstone* pit or *working*.

STONE TUBBING. Water-tight stone *walling* of a *shaft*, jointed and fastened at the back with cement.

STONE WORK (S.). Driving of drifts or galleries in *measures*. See *Stonehead* (1).

STOOK (N.). A pillar of coal about four yards square, being the last portion of a full-sized *pillar* to be worked away in *board and pillar* workings.

STOOK AND FEATHER. A wedge for breaking down coal, worked by hydraulic power, the pressure being applied at the extreme inner end of the drilled hole.

STOOL (D.). To *sit*, which see.

STOOLS (F. D.). *Sigillariæ*, viz. the fossil form of the stem of a tree, which grew during the Coal period, occasionally met with (probably *in situ*) in mines.

STOOP. 1. (S.) See *Rance*.

2. (M.) A *prop* or *puncheon*.

STOOP AND ROOM (S.). A system of working coal very similar to *pillar and stall* (Fig. 122).

Fig. 122.

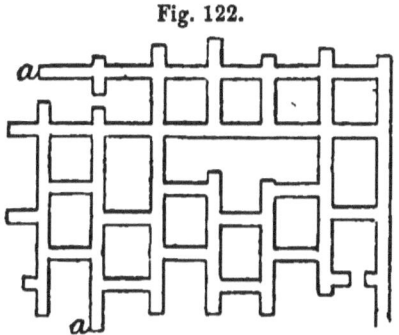

STOOPING (S.). Working away the *stoops* (1).

STOOP ROADS (S.). Roads driven in the solid or *whole* coal on the *stoop and room* system.

STOPPAGES. Deductions from miners' wages, such as rent, candles, blacksmith's work, field club, &c.

STOPPER (S. S.). See *Stopping*.

STOPPING. A solid stone, brick, or clay wall built right across a *thirl* or any other description of *road* or entrance to a worked-out place. They prevent the access of air to *goaves*, and cause it to circulate through and further into the mine; are often plastered with lime on the *intake* side and packed at the back with sand, slack, *burnt stuff*, or rubbish. See *Dam, Signs*.

STOP TRUCKS (S. W.). Scotches.

STOW. To pack away rubbish into *goaves*, old *roads*, &c.

STOW-BOARD (N.). A *board* or *heading* in which débris is *stowed*.

STOWSES (N.). A windlass or *wallow*.

STRAIGHT BIT. A flat or ordinary chisel for *boring*.

STRAIGHT COAL (S. S.). An excavation made in the Thick coal, having the solid coal left on three sides of it.

STRAIGHT ENDS AND WALLS (N. W.). A system of working coal somewhat similar to *board and pillar*. *Straight ends* are *drifts* or *headings* from 4 feet 6 inches to 6 feet in width. *Walls* are *pillars* 30 feet wide.

STRAIGHT WORK or STRAIT WORK. The system of getting coal by *headings* or *narrow work*. See *Course*, Fig. 44.

STRAPS (M.). Old iron *way* rails put up between the coal *face* and the front rank of props, in *long-wall* stalls, for supporting a tender *roof*.

STRAW. A fine straw filled with gunpowder, and used as a fuse.

STREBBAU (Pr.). The *long-wall* system, which see.

STRET. 1. (N. S.) See *Straight Work*.

2. (M.) Solid, close, compact: e. g. *gobbed stret*, *packed stret*, &c.

STRETCHER (Y.). A *prop* or *sprag*.

STRIKE. 1. The line at right angles to the *dip* (3); a level course.

2. To meet with, or *hit* a *fault, hollows,* &c.

STRIKE JOINTS (U. S. A.). Joints in strata parallel to the *strike* (1).

STRIKING DEALS. Planks fixed in a sloping direction just within the mouth of a *shaft*, to guide the *bowk* to the surface.

STRIP (M.) To *get* coal, &c., alongside a *fault*, *barrier, hollows,* &c.

STRIPPING (Y.). A *web* of coal worked off all along the *face* of a *stall*.

STRONG. A word having reference to the character of a *bind* or *metal*, meaning that the argillaceous is largely mixed with the arenaceous or siliceous material.

STRUCK (N.). Level full; strickle measure.

STRUM (N.). A kind of iron sieve placed round the suction pipe of a pump, for preventing stones or other rubbish passing into the pump.

STRUVE VENTILATOR. A pneumatic apparatus invented by a Mr. Struve, consisting of two vessels, something like gas-holders, which are moved up and down in water. By this means the air is sucked out of the mine as required. See *Ventilator*.

STUFF. 1. Coals and slack, the produce of the mine. 2. (Sh.) See *Bind*.

STUMP (Pa.). The block of solid coal at the entrance to a *breast*, having a narrow roadway on either side.

STUMPING (L.). A kind of *pillar and stall* plan of getting coal.

STYTHE. Carbonic acid gas. A gas commonly given off from old workings, and one found to result from the breathing of men and horses, the burning of candles and lamps, and from the explosion of gunpowder used in blasting. Shallow and badly ventilated mines produce *stythe*. See *After-damp* and *Black-damp*.

SUB (M.). Meaning subsist; money or wages paid on account.

SUCK. See *Back-lash*.

SULPHUR (S. S.). Old term for *fire-damp*, which see.

SUMP or SUMPH. 1. The bottom of *shaft* below the lowest *inset*.

2. A portion of the *shaft* bottom of a *sinking pit* sunk down lower than the other, forming a kind of dish into which the water collects, and which is always allowed to be the deepest part.

3. (N.) A portion of a length of a *broken* working, or of a *jud*.

SUMPER. A *shot* placed in or very near to the centre of the bottom of a *sinking pit*.

SUMPT (S. S.). See *Sump*.

SURFEIT (N.). *Choke-damp*.

SURGE. To slip accidentally.

SWABSTICK. A short wooden rod bruised into a kind of stumpy brush at one end, for cleaning out a drilled hole.

SWAD. See *Dant*.

SWAG (L.). Subsidence or *weighting* of the *roof*.

SWALLOW HOLES (L.). See *Sinks*.

SWAMP. A depression or natural hollow in a *seam*. See *Lum*.

SWAPE (N.). A great oar by which *keels* are steered.

SWAYING OF A BANK (Y.). An expression commonly made use of in South Yorkshire, which means that a *bank* (4) is undergoing disturbance in the *roof*, due to *weight* (1, 2).

SWEAL. 1. See *Gutter*.

2. A candle is said to *sweal* when the grease runs down, owing to its burning in a strong current of air or being improperly carried or fixed.

SWEAT (M.). The roof of a mine is said to *sweat* when drops of water are formed upon it, due to the heating of the *waste* or *goaf*. *Sweating* is generally the first indication of a *fire-stink*.

SWEEP-HEAD PICK. A *pick* the form of the head of which is made curved instead of *elbowed* or *anchored*, as other kinds are termed.

SWEET. Free from *fire-damp* or other gases, or from *fire-stink*.

SWELL. A kind of *fault*. See *Horses*.

SWELLY, also SWALLY, also SWILLY (N.). A thickening out of a seam of coal over a limited area.

SWILLIES (Y.). Detached portions of coal strata forming small basins,—say not more than one mile in diameter.

SWINE-BACK (S. W.). See *Horses*.

SWING. The arc or curve described by the point of a *pick* or *maundril* when being used by a *holer* or in cutting coal; called the *swing of the pick*.

SWINGING BONT or BANT (M.). Before the introduction of *cages* and *conductors*, the *skips* of coal, &c., and men were raised and lowered swinging loose in the *shafts*. Very shallow mines are still worked in this manner. The word *bont* means *band*, a rope or chain.

SWOM STUFF. An old term for certain alluvial deposits met with in *coal measures*.

SIPHON or SIPHON-PIPE. A simple, very effective, and economical mode of conveying water in a mine over a hill, or from one *lodge* to another, from a higher to a lower level. It takes the form of an iron pipe (w. i. tubes are perhaps the most suitable), the vertical height of which must not exceed 28 or 30 feet between the water to be run off and the summit of the hill, and the length of the discharge end must exceed in height that of the suction end, or the *siphon* will not work.

# T.

TACK. 1. (N.) See *Spurns*.

2. (Som.) A wooden scaffold put into a *pit-shaft* for temporary purposes.

TACKLE. The ropes, chains, *detaching hooks*, *cages* or *kibbles*, and other apparatus for raising coal, &c., in *pit-shafts*.

TACKLERS or TUCKLERS (Lei.). Small chains put round loaded *corves*, to keep the coal from falling off.

TACKLER SKIP (S. S.). A kind of box in which men used to ride in a *shaft*, used also for carrying minerals. See *Paddy Pan*.

TACKS (N.). The rock walls or sides surrounding a number of *boreholes* (2) which in driving *stone heading* (1) in *fiery mines* are drilled in the *head-end* or *face*, and the *tacks* between them are forced out or cut away without resorting to blasting.

TACKSMAN (S.). The lessee of a colliery.

TAGUE. An iron plate fitted on one side with a semicircular projection or rib, and two other short curved pieces, suited to the gauge of the tram rails, by which the wheels of the trams are guided from the plate on to the rails. See Fig. 123.

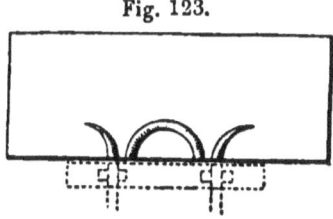

Fig. 123.

TAIL BACK. When *fire-damp* ignites at a *furnace* or by other means, and the flame is elongated or creeps backwards against the current of *air*, and possibly causes an *explosion* of a large body of *gas*, it is said to *tail back* into the *workings*.

TAIL CRAB. A crab for overhauling and belaying the *tail rope* (3) in pumping gear.

TAIL IN (M.). To run out or terminate a length of *holing stints* at a *buttock* or other particular point along the *stall face*, or (if commencing to *open-off stalls*) from the side of a *heading*.

TAILLES CHANSANTES (F.). Coal workings where the strata lie horizontal or nearly so.

TAILLES MONTANTES (F.). Workings to the *rise* or in *steep seams*.

TAIL-PIPE. The suction of a pump.

TAIL ROPE. 1. A round steel or iron wire rope working in conjunction with, and being an appendage to, a *main rope* in the system of underground *haulage*, where the inclination of the *ways* is only slight. By the *tail rope* the empty *set* is drawn *inbye*. They are much used in branch *dip-ways* or slants, in which system they are drawn *inbye* by the weight of the *empties* or by horses, engine-power of course being applied to bring the full *set* back, or *outbye*.

2. A round wire rope attached to *cages* as a balance. See *Köepe System*.

3. A round hemp rope used for moving pumps in *shafts*.

TAIL-ROPE SYSTEM OF HAULAGE. This is worked with a single road or line of rails, and generally applied under the following circumstances. When the average gradient of the *wagon-way* is not sufficient to cause the empty *set* to draw a single rope in after it; when the gradient dipping *outbye* is not sufficient to establish a self-acting *inclined-plane* system; or when the gradient for the full *tubs* is insufficient to enable the *train* to draw a single rope after it. The full set is drawn *outbye* with a *main rope*, and the empty set is hauled *inbye* with a *tail rope*, both ends of the set being attached to a rope. The engine has two *drums*, one for each rope, one always running loose whilst the other is in gear. The *tail rope* is carried upon small sheaves or *rollers*, either on the *floor* or towards the *roof*. The speed of the *set* does not usually exceed 8 or 10 miles per hour.

TAKE. 1. The extent or area of a lease of mineral property—often several thousand acres.

2. (L.) To *show* or reveal *gas*.

TAKE OUT (C.). To *crop* out.

TAKER-OFF (Y.). See *Puller-off*.

TAKE THE AIR. To make experiments with the *anemometer*, or by other means to ascertain the amount of *ventilation* passing through a mine. See *Water Gauge*.

TAKING. A *Take*.

TAKING OF PROPS (L.). *Drawing* the timber in the *wastes* of workings.

TALE (Som.). A day's work or a day's *output* of coal.

TALLY. A mark or number placed by a *collier* (1) upon every *tub* of coals loaded and sent out of his *working place*. They are usually little bits of tin having a number stamped upon them, and hung upon the tub by a short piece of string. By counting the number of these *tallies* when taken off the tubs at surface, and ascertaining the average weight of coal in each tub, the quantity of coals sent out of each *stall* is arrived at.

TALLY-SHOUTER. One who shouts out the numbers on the *tallies* to the weigher.

TAMP. To fill up a *borehole* (2) above the charge with some strongly-resistant substance, such as shale or dirt pounded up small, and rammed hard upon the powder before firing off the *shot*.

TAMPING. The stuff used to *tamp* with. See *Stemming*.

TANGERS (S. W.). Timbers fixed in a particular manner for supporting the sides of *headings* in shifting or very soft ground.

TAP. 1. To cut or bore into old *workings* for the purpose of liberating accumulations of *gas* or *water*.

2. To *win* coal in a new district.

TAPPING THE HOLLOWS. A common expression, meaning allowing *water* or *gas* or both to flow out of disused *workings* (often under a great pressure); an operation requiring great caution, and occasionally attended with risk.

T Chisel. A boring tool with its cutting edge made in the form of the letter T, but a little curved, T.

Teem, sometimes Tem. To tib rubbish, &c., down a *spoil-bank*. See *Dump*.

Teeming Trough (L.). A cistern into which the water is pumped from a mine.

Teeth-work (S.). Signifies working coal *end on*, which see.

Telegraphs (Pa.). Shoots which convey coal from *screens* (1) to pockets at *breakers*.

Temper Screw (Pa.). See *Stirrup*.

Ten (N.). A certain weight of coal agreed upon between lessor and lessee, upon which a *royalty* is paid at so much per *ten* of *round* and so much per *ten* of *small*. A *ten* varies between 48 and 50 tons, or 18½ Newcastle chaldrons of 53 cwts.

Tentail Rent (N.) A rent or *royalty* paid by a lessee upon every *ten* of coals which are worked in excess of the *minimum* or certain rent.

Tenter. A man who has the control or working of an engine or *jig*, or who looks after the horses in a pit.

Thick Coals or Thick Seams. *Coal seams* of greater thickness than (say) 8 or 10 feet (sometimes met with as much as 130 feet), or those which are worked in two or more stages or *lifts* (3). The *Thick coal* of South Staffordshire is about 28 or 30 feet thick.

THICKNESS (of a *fault*). It is measured by the line *a b* (Fig. 77). See *Hade*.

THILL (N.). See *Floor*.

THIN OUT. A coal or other seam of mineral is said to *thin out* when it decreases in thickness so as to become unworkable at a profit.

THING. 1. (N. S.) A straight *facing* from *floor* to *roof*, and often many yards in length.

2. (M.) A *fault slip*.

THIN SEAMS, THIN COAL. Coal *seams* (say) less than 3 feet in thickness.

THIRL or THIRLING. Sometimes *Thol* and *Thurl*.
1. See *Cross-hole*.
2. (Lei.) To cut away the last *web* of coals, &c., separating two *headings* or other *workings*.

THREAD. 1. (M.) See *Cleat*.

2. (M.) A more or less straight line of *stall faces*, having no *cuttings, loose ends,* or *fast ends* or *steps*.

THROUGH AND THROUGH (S. W.). The system of getting or cutting bituminous coals without regard to the size of the lumps.

THROUGH COAL (S. W.). See *Altogether Coal*.

THROUGHER (S.). A *thirl* (1) put through between two *headings* which are *up-stoop*.

THROW. 1. (Y.) A *fault* of dislocation.

2. The vertical distance between the two fractured ends of a bed of coal, &c., at a *fault*. See *Hade*.

THROWN. Faulted, broken up by a *fault*.

THRUST. *Creep* due to *weight*. When the *floor* is

harder than the *roof*, the subsidence of the latter causes a crushing down of *pillars*.

THWARTING (Som.). A short *branch* (1) driven between two or more *veins* where they are nearly vertical.

TIE-BACK. A beam serving a similar purpose as a *fend-off* beam, but fixed at the opposite side of the *shaft* or inclined road.

TIGER. See *Nipping Fork*.

TIGES DE SONDAGE (F.). *Boring rods*.

TILL (I.). *Shale*.

TILLER. See *Bracehead*, but made in a rather different form, and usually of iron.

TIMBER. 1. Pitwood, e. g. *Props, bars, sprags, lagging*, &c.

2. To set, fix, or place *timber* (1) in a mine.

TIMBERER. One who *sets* (5) and *draws props*, puts up *bars* and *lacing* in the roadways and *workings*.

TIME. Hours of work performed by *day men*, labourers, &c.

TIN CAN SAFETY LAMP. A *Davy* lamp placed inside a tin can or cylinder having a glass in front, air-holes near the bottom, and open-topped; thus transforming an instrument of great danger in a rapid current of air into one of great security.

TINKER (D.). Laminated carbonaceous shale.

TIP. A platform upon which a pair of iron tram rails, fixed upon an axle and attached to a lever, are bolted down, for emptying *tubs* into wagons, boats, &c. See Fig. 124.

Fig. 124.

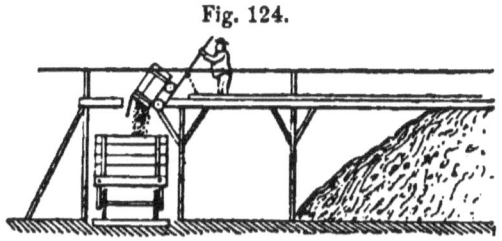

TIPPER or TIPPLER. An apparatus for emptying *tubs* of coal on to *screens* (1). The tub is placed in the *tippler*, turned upside down, and brought back empty to its original position, with a minimum of manual labour. It is constructed principally of wrought iron, and usually fitted with a brake. See Fig. 125.

Fig. 125.

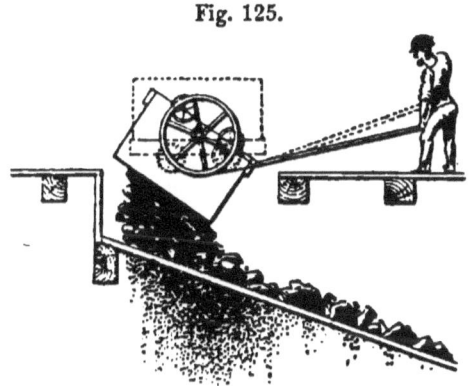

TIRR. See *Overburden*.

TOE. See *Spurn*.

TOKEN. 1. (S. W.) A thin bed of coal, &c., indicating a thicker seam at no great distance.

2. See *Tally*.

TOLL (Ch.). *Royalty* on rock salt.

TOOM (N.). Empty.

TOP. 1. See *Roof*.
2. See *Cap* (1) or *Blue Cap*.

TOP HEADS (S. S.). Passages driven in the upper part of the Thick coal for draining off the *gas*; first adopted by one James Ryan about the year 1808.

TOPIT. A kind of *bracehead*, but much smaller, which is screwed on to the top of *boring rods* when withdrawing them from the *hole* (2). It is attached to a rope worked from a *jack-roll*.

TOPPLE (S. W.) from TOP-HOLE. A *working place* driven to the *rise* of the main levels.

TOPPLY (S.). The uppermost layers of a bed of coal left for a *roof*.

TOPS. See *Top*.

TORRENTS. Beds of quicksand met with below the chalk marl in the Anzin coal-field, in France.

TOT (N.). A measure of gunpowder used in blasting.

TOUCH. See *Fuze*.

TOUGH (Sh.). Grey, plastic clay.

TOUT VENANT (Belg.). Coal as landed on *bank* (1), previous to *screening* (1) and sorting.

TOW. 1. (Lei.) Dark, tough, earthy clay or shale.
2. (S.) A *winding rope* of hemp.

TRACK (Pa.). Underground railways or tramways.

TRAILER (N.). See *Putter*.

TRAIN. See *Journey*.

TRAIN BOATS (Y.) A number of compartments hinged together in a simple manner admitting of free articulation, in which coals are carried on canals or rivers from the mines to the shipping ports. The *train* may either be propelled or towed. When towed, as many as 30 compartments are linked together, but when propelled the *train* consists of 10 compartments steered by means of wire ropes along the sides, these ropes being actuated by steam power. Each compartment has a capacity of from 35 to 40 tons.

TRAIN BOY. A lad who rides upon the *train*, to attend to the rope attachments, signal in case of derailment of *tubs*, &c.

TRAM. 1. See *Box, Corf, Tub, Skep*. In South Wales *trams* constructed wholly of wrought iron or steel are much used in the steam-coal collieries. They weigh about 9 cwt. empty, and have a carrying capacity of 25 cwt. See Fig. 126.

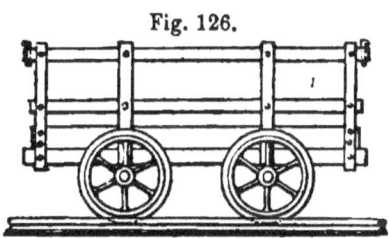

Fig. 126.

2. To haul or push *trams* (1) about in a mine.

TRAMMER. See *Haulier, Putter*.

TRAMMING. See *Haulage*.

TRAM-PLATE. Cast-iron plates of L section, weighing about 12 lbs. to the yard, upon which wagons and *trams* run. See *Tram-road*.

S

TRAM-ROAD. A road laid with tram rails or plates. So called after one Benjamin Outram, of Little Eton, in Derbyshire, who in 1800 used stones for carrying the ends of the metal plates or edge rails. The name Outram was subsequently contracted into Tram, hence tramway, trams, &c.

TRAM-ROPE. A hauling-rope to which the *trams* are attached by a *clip* or chain, either singly or in sets. Round steel ropes are always used.

TRAP. 1. (S.) A steep *heading* along which men travel.

2. (B.) See *Lid*.

3. (Som.) A *fault* of dislocation.

4. See *Grappel*.

5. See *Whin*.

TRAP DOOR. A small *door*, kept locked, fixed in a *stopping* or *bolt*, for giving access to *firemen* and certain others to the *return air-ways*, *dams*, or other disused places in a mine.

TRAP-DOWN (B.). A *fault* which is a *down-throw* one.

TRAP DYKE. A *fault* (not necessarily accompanied by a displacement of the strata) in which the spaces between the fractured edges of the beds are filled up by a thick wall of igneous rock called *trap* (5) or *whin*. Frequently met with in the collieries of the North of England and Scotland. The word *Trap* is derived from the Swedish *Trappa*, a stair.

TRAPPER (N.). A small boy employed underground

in opening and shutting *doors* during the passage of *tubs* and horses.

TRAPS (S.). Travelling roads for miners in *Edge Coals* driven on the slope of the *seam*.

TRAP-UP (B.). A *fault* which is an *up-throw* one.

TRAUNTER (M.). A *sprag*. See *Tront*.

TRAVAIL À COL TORDU. (F.). See *Holing*.

TRAVELLING ROAD. An underground passage or *way* used expressly, though not always exclusively, for men to travel along to and from their *working places*.

TREE. 1. See *Leg* (1), *Puncheon*.
2. A *pump-tree*, which see.

TREE UP (S.). To set up *props* or *trees* (1) in the *workings*.

TRÈPAN. 1. (F.) A boring chisel of the ordinary form.

2. The *boring head* or tool used in the *Kind-Chaudron* system of *sinking shafts*. It consists essentially of a horizontal wrought-iron bar, to the underside of which are attached steeled teeth, so placed, that as the bar is rotated round the central axis of the pit, each tooth in falling with the bar through the requisite length of the stroke, which is from 10 to 20 inches, cuts for itself an annular portion of the bottom of the *shaft*. A large and a small *trèpan* are used: the smaller one first bores out a hole from 4 to 6 feet in diameter, according to the required size of the *shaft*, in advance of the full size of the pit, into which the débris falls. The *trèpans* are suspended by long wooden rods, and for a *shaft* of a diameter of say 15 feet, the larger one will weigh

about 20 tons, and the smaller say 11 tons. In ordinary strata the average daily advance of the boring will be about 3 feet. Fig. 127 is a large trèpan.

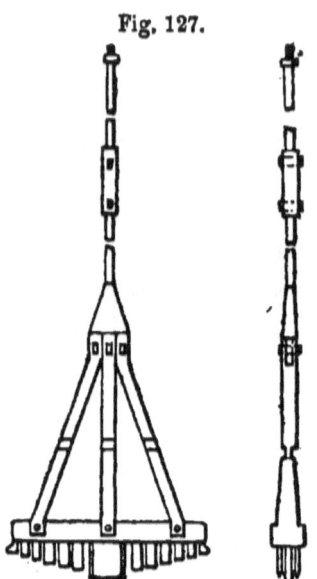

Fig. 127.

TRIG. A *sprag* used for stopping or putting the brake on *trams, wagons*, &c.

TRIMMER. See *Pricker* (3).

TRIMMERS (N., S. W.). Men who fill up the holds of vessels (*colliers* (2)) with the coals discharged into them from *staithes*.

TRIP. See *Kick-up* or *Tipper*.

TRIPLET (N.). See *Tipper*.

TROLLEY. 1. A *Tram*.

2. (B.) A kind of *Lum*, or basin-shaped depression in strata.

TROMMEL. To separate coal into various sizes by discharging them with the least possible breakage.

TROMPE. A water-blast apparatus for producing *ventilation* by the fall of water down a *pit-shaft*. It consisted of a pipe, which the water enters in a funnel-shaped stream, and regulates the discharge of water; the air enters chiefly through holes just below; the water breaking on a block is forced through the air-pipe or trunk.

TRONT (M.). A long *sprag* fixed diagonally to the *face* of the coal *wall*.

TROUBLE. A *Fault*.

TROW (Lei.). A rectangular wooden pipe made in lengths of 12 or 14 feet, and from 3 to 12 inches square inside, for conveying the water *feeders* down the side of a *shaft* to the *garlands* (1). Used also occasionally for ventilating a trial *heading, staple*, or other nook-and-corner in the *workings*.

TROUSSE COLLETÉE (F.). A narrow *wedging crib* placed beneath an ordinary one.

TROUSSE PICOTÉE. An ordinary *wedging crib*.

TROUGH FAULT. A wedge-shaped *fault*, or, more correctly, a mass of rock, coal, &c., let down in between two faults, which faults, however, are not necessarily of equal *throw* (2). See Fig. 128.

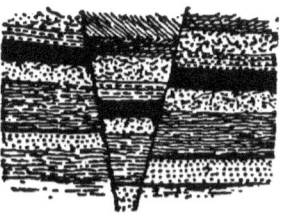

Fig. 128.

TRUCK. See *Tram*.

TRUMPET LAMP (N.). Miner's term for a *Mueseler* or Belgian safety-lamp.

TRUMPETING (S. S.). See *Brattice*. Fig. 27 brick.

TRUNCHEON (Som.). A sleeper for underground railways.

TRUNK. 1. (M.) A wooden box or sledge or *sled* in which the débris is conveyed from a *heading* of very small sectional area, or up a *staple*.

2. (B.) A wooden pipe or box for conveying *air* in the *workings*.

3. (Y.) See *Kibble*.

TRUNK PUMPING-ENGINE. One which commands the

drainage of underground waters over a considerable area of mines, being a substitute for a number of smaller and independent pumps.

TRUNT (N. S.). A *heading* driven on a level.

TRYING THE LAMP. The examination of the flame of a *safety lamp* for the purpose of forming a judgment as to the quantity of *fire-damp* mixed with the *air*. When *fire-damp* forms 1 part out of 13 of air, the mixture becomes explosive; when 9 to 10 parts of air to 1 of *gas*, the explosive force is greatest: 5 parts of air to 1 of *gas* causes the most feeble explosion.

TUB. 1. See *Box, Corf, Tram*.

2. A complete length of metal or timber *tubbing* from and including the *wedging crib* upwards.

TUBBED BACK. Springs or *feeders* of water met with in *sinking pit-shafts* are said to be *tubbed back* when *tubbing* has been put in to keep the water from getting into the mine.

TUBBING. Cast-iron and sometimes timber lining or *walling* of a *pit-shaft* to keep back springs of water from flowing into a mine. See *Plank tubbing*. Of metal tubbing there are three kinds employed, viz.—

1. Ordinary outside-flanged tubbing, put in in segments and wedged up water-tight.

2. Inside screwed tubbing put in in *rings* (1) and segments bolted together and wedged, either built up from a *wedging crib* or lowered from the surface as a cylinder through water-bearing strata to the *stone-head* (2).

3. Complete rings or cylinders built up one above

another at surface as they are lowered into the pit, bolted together at the joints, which have inside flanges. See Fig. 129, showing the three systems in plan as well as in section.

Cast-iron *tubbing* first used in 1792, at Wallsend.

Fig. 129.

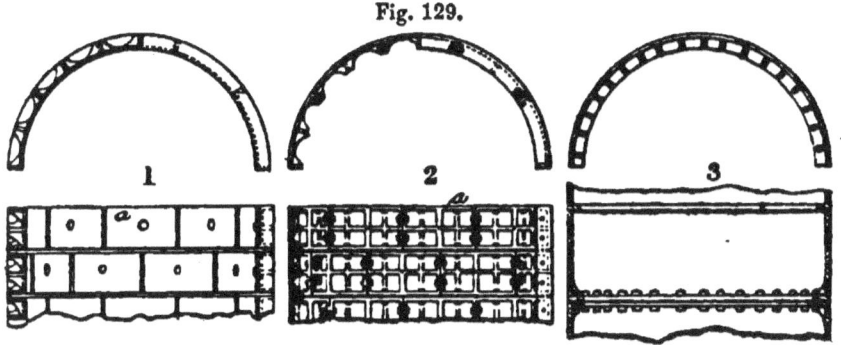

TUBBING PLATES. Cast-iron segments forming portion of a ring of *tubbing*. See Fig. 129, 1 and 2, *a a*; also enlarged views, Fig. 130. Generally from 10 to 12

Fig. 130.

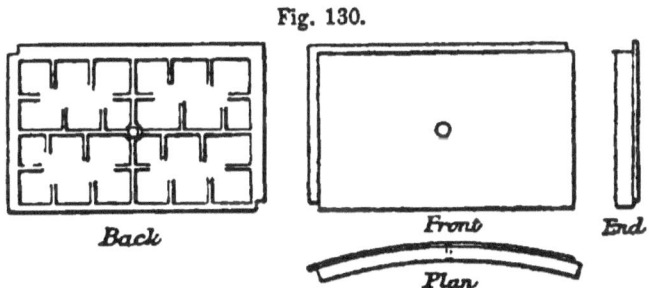

plates form a *ring* (1). Thickness of the metal from 1 to 2 inches, according to the pressure of water.

TUBBING WEDGES. Small wooden wedges of pitch pine about 4 inches in length, 1½ inches in width, and ¼ inch in thickness at the thick end. They are hammered in between the joints of *tubbing plates* until no more can be made to enter, thus stopping back every drop of water from the *shaft*.

TUBING. The lining of *boreholes* (1) with wrought-iron tubes to keep the sides from running in.

TUB WAGON (L.).

TUB-WAY (N.). Tram-rails, sleepers, &c.

TUCKLERS (Lei.). Short chains formerly used for raising and lowering men in a *shaft*. Three men generally sat in them at one time. See *Bant, Tacklers*.

TUGGER (B.). A short chain by which boys draw *tubs* along.

TUGGER BOY (B.). One who draws small *tubs* (1) or *sleds* about underground by means of a *tugger*. Called *Tugger-work*.

TUMBLER. 1. (N.) A stop, scotch, or catch, affixed to each *deck* of a *cage* for keeping the *tubs* in place.

2. (S.) See *Tipper*.

3. (Som.) See *Kneeler*.

TUMBLING TOMS. *Tippers* that turn completely over.

TUNNA (Sw.). See *Bowk, Kibble*.

TUNNEL (L.). See *Crut*.

TURN. 1. The hours during which coals, &c., are being raised from the mine.

2. See *Shift*.

3. To draw or *wind* (3) coals up a *shaft* or up an inclined plane to the surface.

4. Curved *tram* rails laid round a corner or turn, often made of cast iron.

5. (S.)  To drive *headings* to form *stoops*.

TURN AGAIN (N. S.). A change in the direction of the *dip* of the strata.

TURN BARREL (M.). See *Jack-roll*.

TURNING. Drilling a shot-hole by hand.

TURNING OUT (S. S.). Bringing coals to the *skips* (1).

TURN OUT. A siding or pass-by upon an underground *rolley-way*.

TURN PULLEY (M.). A pulley wheel fixed at the *inbye* end of an *endless* or *tail-rope* hauling *plane*, round which the rope returns. It may be fixed either vertically or horizontally, and is usually from 4 to 6 feet in diameter. See *Lurry* (1), Fig. 94.

TURN-STAKES. See *Stowses*.

TURNTABLE. A cast-iron disc or small horizontal platform revolving on a vertical axis, and supported upon small wheels, upon which *tubs* or *trams* are turned round upon the *pit bank*.

TWIBILL. A strong *pick* used for *stone-work*, with an eye generally rectangular.

TWIN BOY (B.). A small boy employed underground to push *trams* along a *twin-way*.

Fig. 131.

TWIN-WAY (B.). Two branch roads set away, one on either side, out of a *main road* to the *face* of the *stalls*, through which *trams* are pushed by *twin boys*. See plan, Fig. 131.

Two (S.). A *cage*-ful of men.

Two-throws. When in *sinking*, a depth of about 12 feet has been reached, and the débris has to be raised to surface by two lifts or throws with the shovel (one man working above another). At this point the employment of a hand windlass becomes necessary.

Tymp. See *Cap* (2), *Lid*. Usually about 12 or 15 inches in length.

Types (S.). See *Lypes*.

# U.

U. C. *Upcast* shaft.

Udged (D.). Loose, weak, liable to fall, sounding hollow, or unsound. A *roof* or a piece of *side* is said to *knock udged* when it produces a dead, hollow, unsafe sound, upon being knocked upon with a hammer, &c.

Umbrella. See *Bonnet*.

Undercast. An *air course* or *wind road* carried underneath a *wagon way* or other road by constructing a kind of bridge made airtight, or by driving a *heading*

Fig. 132.

in solid coal, &c., beneath the *floor*, *sinking* or sloping down at either end. See Fig. 132.

Underclay. A bed of *fireclay*, *clunch*, or other

more or less clayey stratum lying immediately beneath a seam of coal, and met with as forming the *floor* of almost every bed of coal. Many geologists consider *underclays* to have been the soil or surface upon which the vegetation, now converted into coal, grew, flourished, and died, as they contain the fossil remains of great numbers of what are thought to have been the roots of plants, &c.

UNDERCLIFF (S. W.). Argillaceous shale forming the *floor* of many coal seams in this coal-field.

UNDERCUT. To *hole* (1) or *kirve*.

UNDEREARTH (F. D.). A hard bastard *fireclay* forming the *floor* of a seam of coal.

UNDEREDGE STONE (F. D.). The floor of an *ironstone* mine.

UNDER-GETTINGS. See *Shorts* (2).

UNDERGOING. See *Holing, Kirving*.

UNDER-LEVEL (Cl.). *Winning* (1) the ironstone by driving *drifts* into the hill-sides, &c., instead of *sinking shafts*.

UNDERLOOKER (L.). One who has the care and superintendence of the colliers or miners and of the *workings*, who receives his orders from the *manager*, and to whom the *overmen* and *deputies* report upon the state of the mine.

UNDERPINNING. Building up the *walling* of a *pit-shaft* to join that above it.

UNDERPLY (S.). A band or division of the upper portion of a thick seam of coal.

UNDER-ROPE. See *S-rope*.

UNDER-SEAMS (S.). Lower or deeper coal seams.

UNDER VENTILATION. Too little *air* circulating in a mine or *working-place* therein.

UNDERVIEWER (N.). See *Underlooker*.

UNGOTTEN. See *Unwrought*.

UNHOLED (Y.). *Boardgates* or other *headings* which are not *driven* through or *thirled* into the adjoining roadway.

UNWATER. To pump mines, or *districts* in mines, dry.

UNWROUGHT or UNWORKED. Coal or other mineral which has not been mined or worked away.

UP. 1. A *stall* or *heading* is said to be *up* when it is driven or worked up to a certain line (a *fault, hollows*, boundary, &c.), beyond which nothing further is to be worked.

2. On the *bank* (1) or on the surface.

UP-BROW (L.). An inclined plane worked to the *rise*.

UPCAST. The *pit-shaft* through which the *return air* ascends and is got rid of. See *Signs*.

UP-HILL. A *board* or *wicket*.

UP-LEAP (M.). A *fault* which appears as an *upthrow*. See *Fault*, Fig. 70 (1). From *c* to *d* is an *up-leap*.

UP-OVER CRIB. A *wedging crib* placed on the top of a length of *tubbing*, to *tub* (3) off the water in a certain stratum.

UPSET (S.). A *bolt hole* or *thirl* (1) put through between two levels in *edge coals*.

UP-STANDING (N.). The condition of a *goaf* when such portions of the *pillars* are worked away as still to leave the *roof* supported.

UP-STOOP (S.). When a *heading* is driven to a point at which another should be put in or meet it at right angles out of a parallel *heading* so as to form a *stoop*, the first-named *heading* is called *up-stoop*. The headings or *rooms* marked with the letter *a* in Fig. 122 (see *Stoop and Room*) are *up-stoop*.

UP-THROW. See *Up-leap*.

## V.

VACUUM. The method of producing *ventilation* by exhausting the air from the mine. See *Fan*.

VEAL (S.). A tank or water-barrel placed upon a *cage* for emptying the *sump*.

VEE (M.). The junction of two underground roadways meeting in the form of a V.

VEERER (Som.). An old word for *Banksman*.

VEES, VEEZ, and VIESE (S.). A kind of soft earth in a fissure or upon the sides of a *dyke*. See also *Leather-bed*.

VEIN (S. W.). A seam of coal.

VEISES (S.). Joints in the coal strata.

VENT or VENT HOLE. A small passage made with a *needle* through the *tamping*, which is used for admitting a *squib*, to enable the charge to be ignited.

VENTILATING COLUMN. See *Motive Column*.

VENTILATING PRESSURE. The power or force required to overcome the friction of the air in mines. This is found to increase and decrease in exactly the same proportion that the area or extent of the rubbing surface exposed to the air increases or decreases. The rubbing surface depends upon the perimeter of the airways and their length. See *Drag* (1).

VENTILATION. 1. The atmospheric air circulating in a mine.

2. The art or method of producing, distributing, maintaining, conducting, and regulating a constant current or flow of atmospheric air in the *shafts, levels, inclines, staples,* engine- and boiler-houses, stables, *returns,* flues, edges of *goaves,* of old *workings,* &c., so as to dilute, and as far as possible render harmless, the noxious gases given off in the mine, and in that state to convey them into the atmosphere at the surface. See *Natural Ventilation, Furnace, Steam Jet, Fan.*

VENTILATOR. A mechanical apparatus for producing a current of air underground.

There are about ten different types at work, all of them being on the exhausting principle. They may be divided into two clearly and radically distinct classes, the first consisting of the Guibal, Rammel, Waddle, and Schiele Ventilators, which are centrifugal *fans,* and act by reason of the partial vacuum they are able to produce; and the second consisting of machines known as varying-capacity ventilators, and which act in a similar manner to an air-pump. They are known as the Nixon, Struvé, Lemielle, Cooke, Root, and Goffint Ventilators (see Fig. 133, which gives all the abovementioned ventilators in side elevation, with the excep-

Fig. 133.

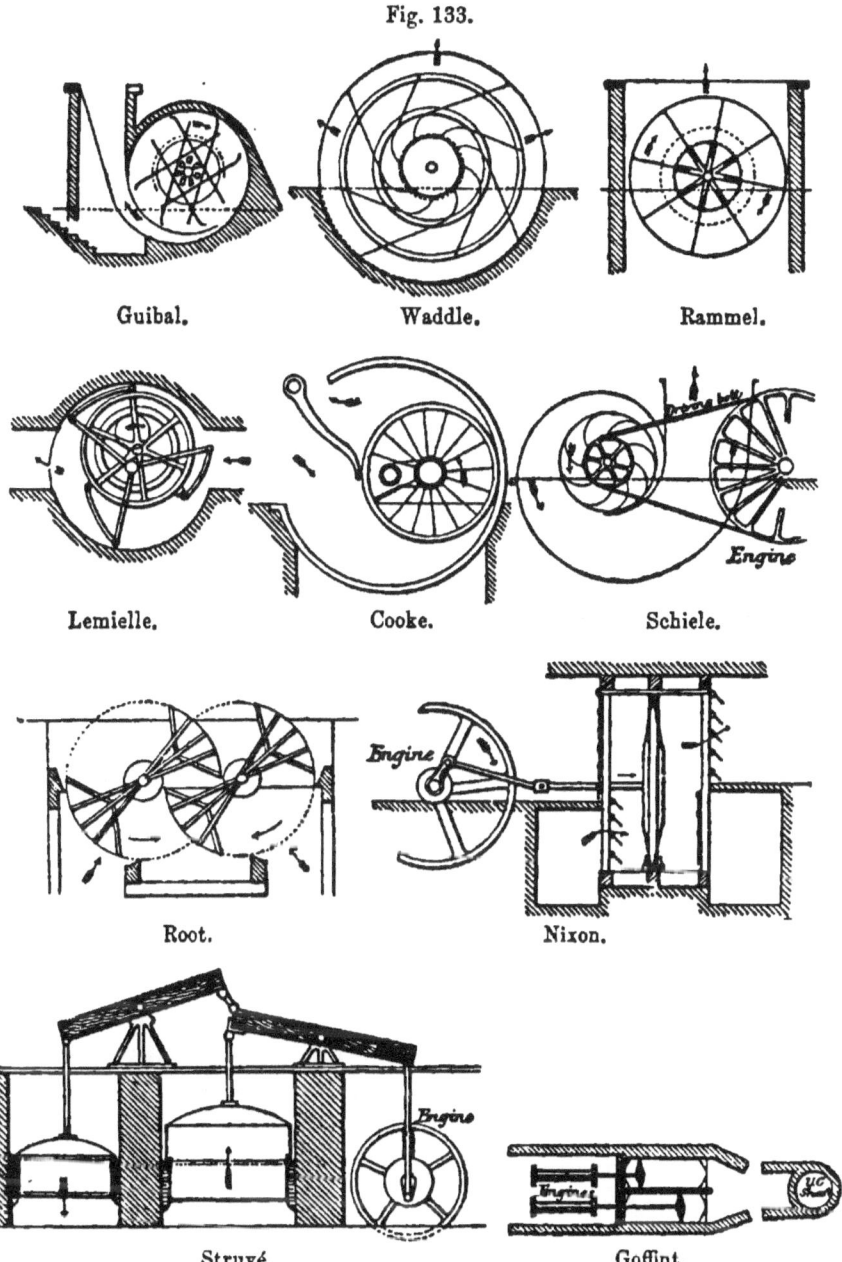

tions of the Lemielle and the Goffint, which are in plan. The centrifugal ventilators are chiefly constructed of wrought iron or of steel, with cast metal central bosses, and are made up to 46 feet in diameter (Schiele up to 14 feet 6 inches). Lemielle's machine consists of a vertical cylinder, within which revolves a second cylinder or drum, also vertical, the axis of which is placed eccentrically to the outer one. Upon this cylinder are hinged doors, which act upon the air in a somewhat similar manner to what the feathering float-boards adopted in steamer paddle-wheels do upon the water.

Cooke's Ventilator consists of two horizontal drums mounted eccentrically upon a shaft: each drum as it revolves moves almost in contact with a cylindrical casing. A vibrating arm or *shutter* is hung by the upper edge, and the lower edge is kept closely in contact with the surface of the revolving eccentric cylinder.

Root's Ventilator is a rotary displacement machine, discharging the air in four distinct volumes during each revolution. It consists of two rotary pistons revolving in a casing. They are constructed of wrought iron and timber, and adjustable packing blocks are provided at each end of the ventilator chamber to prevent slipping of the air.

The Nixon Ventilator consists of an enormous horizontal double-acting air-pump, fitted with rectangular pistons running to and fro upon rails. Upon the fronts and backs of the chambers are hung a number of rectangular valves or *flaps*, through which ingress and egress is given to the air.

Struvé's Machine consists of two vertical air-pistons called *aërometers*, constructed of wrought iron, which reciprocate vertically in annular tanks filled with water. The inlet and outlet of the air is regulated by rectangular valves in much the same way as in the Nixon Ventilator.

The Goffint Ventilator (at Liége, Belgium) consists of a horizontal double-acting piston-pump like that of Nixon, but differing in construction from that machine.

VIEWER or COAL VIEWER. The general manager or mining engineer of one or more collieries, who has control of the whole of the underground works, and also generally of those upon the surface. Underground surveys and plans are generally made and kept up by him, and the *Manager* acts under his authority and directions. A word not much used now, and is giving place to Mining Engineer and *Agent*.

VISETTE (F.). See *Slope* or *Incline*.

VORHAUER (Pr.). This word means "Old man of the *stall*." He corresponds to the *first man* or *butty collier* of English mines.

VUGHY ROCK. A stratum of cellular structure, or one containing many cavities.

## W.

WAD COIL. A tool for readily extracting a pebble or a broken tool from the bottom of a *bore-hole* (2), consisting of two spiral steel blades arranged something like a corkscrew. See *Spiral Worm*.

WAD-HOOK. See *Wad Coil*, *Spiral Worm*.

T

WAFF (S.). See *Brush* (1), *Dadding*.

WAFTING (M.). See *Brush* (1).

WAGEMAN (Lei.). A collier who is paid by the day for performing a fixed amount of work, e. g. *blowing*. See *Blow* (1).

WAGON, sometimes WAGGON. See *Box, Corf, Hutch, Skip* (1), *Tram, Tub* (1).

WAGONER (N. S.). A man or boy who goes with a horse hauling *tubs* underground.

WAGON-WAY (N.). An underground *engine-plane* or *horse-road*.

WAILERS (N.). Boys who pick out the *bats* and other rubbish from coal wagons that have fallen through the *screens* (1) unobserved.

WAITERS-ON. Men employed at the top of a *sinking pit* to work the running platform and steady the *kibbles*, &c.

WALL. 1. The *face* (1) of a *long-wall* working or *stall*, commonly called the *coal-wall*.

2. (N.) A rib of solid coal between two *boards*.

WALL ["To the Wall"] (N.). A term signifying breadth, in reference to the size of *pillars* in the system of working known as *Pillar and Stall*.

WALL BARS. *Prop Wood* usually cut flat to fix against the *roof*, close up to the *working face*, where the *roof* is liable to break along the line of *face* (1).

WALL CUTTING. Cutting, shearing, and blasting off the sides of a *sinking pit*, preparatory to putting in *tubbing, coffering,* or *walling*.

WALLING. 1. The brick or stone lining of *pit-shafts*. See *Steining*.

2. (D.) Stacking or setting up ironstone, &c., in heaps, preparatory to its being measured or weighed off.

WALLING CRIB. Oak *cribs* or curbs upon which *walling* (1) is built. They are put in every 6 to 10 yards, according to the nature of the *measures* being *sunk* through.

WALLING STAGE. A movable wooden scaffold suspended from a crab on the surface, upon which the workmen stand when *walling* (1) and *tubbing* are being put in, in a *shaft*.

WALLOW (M.). See *Stowses*.

WALL PLATE (Pa.). Strong timbers or buntons wedged firmly back against the strata, and forming a kind of *walling* (1) of a *pit-shaft*.

WALLS (S.). Short *working faces* or *stalls* (also headings 6ft. in width) from 12 to 20 yards wide.

WALLSENDS or WALLSEND COALS (N., Y.). Strictly speaking, an excellent description of household coal originally produced at a colliery near Newcastle-upon-Tyne, near to the eastern termination of the great Roman wall, and near the sea. Many first-class house coals are now termed *Wallsends*, though they have no connection with the place of that name.

WANT (S.). A clean rent or fissure in strata unaccompanied by dislocation.

WAPPING (Lei.). A roughly-made rope or band of hemp or spun yarn.

WARGUES (F.). See *Horse-gin* and *Gin*.

WARK-BATCH (Som.). See *Spoil-bank*.

WARNERS. Apparatus consisting of a variety of delicately-constructed machines actuated by chemical, physical, electrical, and mechanical properties, for indicating the presence of small quantities of *fire-damp*, heat, &c., in mines. At present most of these ingenious contrivances are more suited to the laboratory than for practical application underground.

WARNING LAMP. A *safety lamp* fitted with certain delicate apparatus for indicating very small proportions of *fire-damp* in the atmosphere of a mine. As small a quantity as 0·03 per cent. can be by this means determined.

WARP (Y.). Blueish-brown, finely-laminated tough clay with pebbles.

WARRANT (L.). Synonymous with *Clunch*, *Pounson*, &c.

WARREN or WARREN EARTH (L.). See *Bind*, *Clunch*, &c.

WASH (N.). Drift, clay, stones, &c. Probably ancient river courses or glacier grooves which have furrowed and scooped out the surface in past ages. See *Hopes*.

WASH FAULT. A portion of a seam of coal replaced by shale or sandstone. See *Fault*, Fig. 70 (2); also see *Low* (2).

WASHING APPARATUS OR MACHINE. Machinery and appliances erected on the surface at a colliery, generally in connection with coke ovens, for extracting, by washing with water, the impurities mixed with the coal-dust or small slack. The principle upon which the process is performed is that of gravitation or precipitation.

A common form of washing apparatus consists of a series of long, gently-sloping wooden troughs or open-topped, flat-bottomed pipes, with appliances for collecting the washed coal. Streams of water are caused to flow along these troughs, carrying with them the coal-dust, which parts with its impurities (stone, shale,

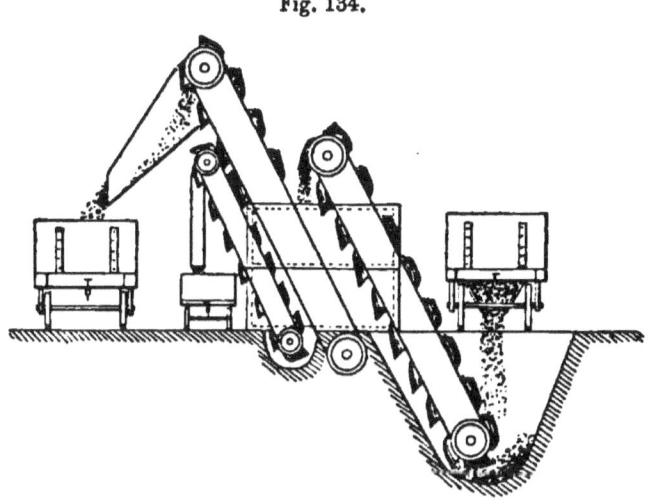

Fig. 134.

&c.), as they soon fall by reason of their greater specific gravity, and the coal passes off into settling-tanks, the water if necessary being pumped back and used over and over again.

Another form of machine, which is much more compact, consists of a brick hopper, constructed below the surface level, into which wagons discharge the coal to be washed. An endless chain of buckets, actuated by an engine, raises the stuff and empties it into iron tanks, wherein the process of cleaning is performed. Out of these a second endless chain of buckets raises the washed and semi-dried coal and tips it over and down a shoot into wagons for removal to coke ovens, a third series of buckets disposing of the washed-out rubbish from the base of the tanks into *trams* or *tubs* for removal to *spoil-bank*. (See Fig. 134). See *Wet Separation*.

WASTE. 1. A more or less empty space between two *packs*. See *Goaf*.

2. (N.) Very small coal or slack.

3. (N.) A *Return Air-way*.

WASTE COAL. Coal obtained from out of a *waste* (1).

WASTEMAN (M.). One who looks after and keeps clean the *airways* of a mine, and keeps the *wax dams* in proper condition. He is generally an oldish collier who has had much experience.

WATCHERS (Lei.). Experienced colliers—*butties*—who take it in turns to go down the pit and examine the whole of the *workings* along with a *deputy* every Sunday.

WATER. Next to *fire-damp*, this is the most troublesome and dangerous element met with in mines. It may, nevertheless, under favourable conditions, be turned to great use in assisting to drain those portions of the *workings* which are situated to the *dip* of the *shafts* or *adits*, through the medium of the *hydraulic pumping-engine* and the *siphon*.

Below a depth of from (say) 900 to 1200 feet it is

seldom found in any quantity, but salt water has been met with at 2790 feet below the surface in a coal-pit. The largest and strongest springs and *feeders* occur within a few hundred feet of the surface, and as many as 12,000 gallons per minute have had to be contended with in *sinking shafts* in the county of Durham.

Brine is occasionally present in coal seams : e. g. at Moira, in Leicestershire, the water pumped from 730 feet in depth out of the "main" coal seam contains no less than 3700 grains of chloride of sodium per gallon. In order to keep water out of *pit-shafts*, several methods of lining them are adopted, viz. *Tubbing, Coffering, Kind-Chaudron* system of *sinking*, and pumping ; and to exclude it from the underground workings and passages a system of *Pillar and Stall* working (which allows a portion of the coal to be extracted, and preserves the *roof* intact, and gives rise to no *weighting* or subsidence of the *cover* containing the water) must either be followed, or it must be raised by pumps or in tanks, or passed off by *adits*.

WATER-BALANCE MACHINE (S.W.). An antiquated method of raising minerals in a *pit-shaft* by water power. The principle of the apparatus consists in a bucket of water, which was filled at the surface, and by its descent raised a *tram* of 20 cwt. or so of coal, the water being run off at the pit bottom each run or *wind* (3).

WATER BARREL. 1. A wrought-iron tank or cistern in which the water is raised from the *sump* or from a *lodge* in the side of the *shaft* by the *winding engine*.

2. An iron or wooden tank or box mounted upon

four wheels, running on the underground tramways, and hauled either by engine power or by horses to the *shaft* bottom, where the water is discharged into the *sump*.

WATER BLAST. The sudden escape of pent-up air in *rise workings* under considerable pressure from a head of water which has accumulated in the lower workings.

WATER CARTRIDGE. Cartridges of explosive substances for blasting down coal in the *workings*. The case containing the powder, tonite, &c., is surrounded by an outer one of water, which is employed to destroy the flame produced when the *shot* is fired, thereby lessening the chance of an *explosion* should *gas* be present in the *place* (1).

WATER CURB. See *Garland* (1).

WATERED. Containing much water—full of springs or *feeders*: e. g. heavily *watered* mines, heavily *watered measures*, &c.

WATER ENGINE (D.). A pumping-engine.

WATER GAUGE. An instrument for measuring the *drag* or friction of air in mines. It generally consists of a glass tube, bent into the form of the letter U, with a scale of inches and parts, by which the difference between the height of the water in one tube and that in the other is measured, this difference being due to the difference of pressure of the air in the *intake* and *return*.

WATER HAMMER. The hammering noise caused by the intermittent escape of *gas* through water in mines.

WATER LEAF (S.). See *Top ply*.

WATER LEVEL. An underground passage or *head* (1) driven very nearly dead-level or on the *strike* (1), for the purpose of draining off the water.

WATER LOAD (S. W.). The head, or pressure per square inch, of a column of water in pumps, &c.

WATER LODGE. See *Lodge*.

WATER-PACKER (Pa.). A kind of cup-leather arrangement fitted to the *tubing* of a *borehole* (1) in watery ground, to keep back the water.

WAX (Lei.). Soft or puddled clay used for *dams* (1) or *stoppings*, and in which the colliers stick and carry about their candles in the mine.

WAX DAM (Lei.). A wall or *dam* (1) of clay.

WAXING (Lei.). The operation of plastering a *waste stack* with *wax*. See *Stack out*.

WAX WALL (Lei.). A clay wall about ten inches in thickness built up from *floor* to *roof* alongside a *gob road* a few feet within the *goaf*, to keep back or prevent *fire-stinks*, &c.

WAY. 1. (N. M.) Any underground passage or *heading* driven more or less on the level of the coal, along which the produce of the mine is conveyed either by horses or by engine power. See *Gate*, *Road* (1), *Wagon-way*.

2. The *rails*, sleepers, chairs, keys, &c., upon which *tubs* or *corves* run.

WAY DIRT (Lei.). The *slack*, *dust* (2), and odd lumps of coal which fall from the *tubs* upon the *roads* on their journey from the *working places* to the *shafts*.

It is collected during the night and sent to *bank* (1), and consumed under the boilers.

WAY END. See *Gate End*. In *long-wall* workings the colliers generally keep a supply of *prop-wood*, a tool and candle box, and other requisites for carrying on their work, and generally take their *bait* or *snap* just within the *way end*.

WAY GATE. See *Gate*.

WAY HEAD (M.). The end of a *way* or *gate* next to the *face*.

WAY LEAVE. 1. A rent or royalty paid by the owner or lessee of a mine for conveying minerals belonging to one person through the property of another person. It is usually fixed at so much per ton, but sometimes, though rarely, depending upon distance conveyed underground and up the shafts.

2. (N.) The right of making and maintaining colliery railways through private property which may intervene between collieries and *staithes*.

WEATHER. To fall or crumble down by exposure to the atmosphere. Certain rocks of the coal measures, such as *fireclay*, *bind*, &c., *weather* very rapidly.

WEB (M.). The *face* (1) or *wall* of a *long-wall stall* in course of being *holed* and broken down for removal. The *web* varies in thickness (according to the height of the *seam*) from 2 or 3 to 7 feet. Fig. 135 shows a cross-section of a *long-*

*wall stall* with a *web* of coals after *drawing* (2) the timber.

WEDDING (D.). The accidental meeting or collision between a loaded and an empty *corf* in a *pit-shaft* working *swinging bont*. Formerly it was not an uncommon thing for the full *corf* or *skip* to come up to surface with the empty *corf* entangled with it.

WEDGING CRIB. A *curb* or *crib* of cast iron upon which *tubbing* is built up and wedged tightly to, in order to stop back all water. *Wedging cribs* are usually about 6 inches thick (though cast hollow), and from 14 to 24 inches broad. More than one are sometimes put in, one on the top of another. See Fig. 136.

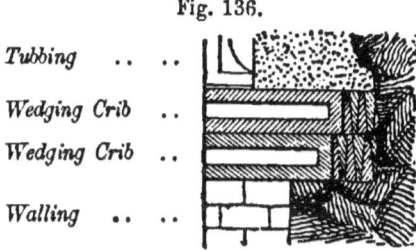

Fig. 136.

Tubbing
Wedging Crib
Wedging Crib
Walling

WEDGING DOWN. Breaking down the coal at the *face* (1) with hammers and wedges instead of by blasting.

WEDGING OUT. Cropping or thinning out. See Fig. 70 (7).

WEDGE RING. See *Wedging Crib*.

WEELDRONS (F. D.). Ancient ironstone workings.

WEEP. See *Bleed*.

WEIGH (S. W.). A weight of 10 tons of coal, &c.

WEIGHER. A man who takes account of the weight

of the contents of every *tub*, or of a certain proportion of the *tubs* of coal, &c., as they leave the *cage* at *bank* (1), or who weighs the coal, &c., in railway wagons, carts, boats, &c.

WEIGHING. The crushing or falling in of the *roof* more or less rapidly.

WEIGHMAN. See *Weigher*.

WEIGHT. 1. A settling or subsidence of the *roof*, due to the working away of the coal seam. *Weights* are commonly of very heavy nature, and make great havoc with the *pit-props* and with the *stalls*.

2. The gradual and regular settlement of the *roof* and *cover*, taking place as the excavation of the seam of coal, &c., goes forward, which by proper management in the working of the coal, and attention to the *goaf*, may generally be utilised in assisting in breaking down the coal in *long-wall faces;* in other words, the *weight* enables the coal when *holed*, to *get* itself. When, in the course of clearing out a considerable area of a seam of coal, &c., and leaving no *posts* or *pillars* of solid coal to support the *roof*, in commencing to *open off workings*, a *weight* (1) takes place. Such *weight* is called the *first weight*, because it is the first crushing down of the *roof*, &c., of any magnitude that has occurred since beginning to form a *goaf*. With *first-weights* generally comes much *firedamp*, as well as much difficulty in keeping the working places safe to work in, owing to *falls*. See *Web*, Fig. 135, showing the serviceable action of *weight* upon a *long-wall working face*.

3. The number of hundredweights (cwts.) which are reckoned as one ton, as between *coal-masters* and

workmen (*hewers, trammers, banksmen,* &c.). In days gone by, as many as 25 to 30 cwt. were allowed to the ton, to compensate for *dirt,* &c., sent out of the pit along with the coal. This was called a *long ton.*

WEIGHTING. Undergoing disturbance due to *weight* (1). Commonly known as being *on the weight.*

WEIZE. A band or ring of spun yarn, rope, gutta-percha, lead, &c., put in between the flanges of pipes before bolting them together, in order to make a watertight joint.

WET SEPARATION. The various systems of cleaning coal at surface by washing, the principle of them consisting in that the various fragments of shale or *dirt* (1) are, by reason of their specific gravity, effectually separated from the coal.

WETTERAUFSEHER (Pr.). A man set aside for the special purpose of attending to the *ventilation.* He carefully examines the mine before the other workmen enter, and reports himself to the *steiger.*

WETTERMAN (Pr.). A trustworthy *collier* (1), who is head man in a *stall* or other *working place.*

WETTER SOHLE (Pr.). See *Air Level.*

WEY. A certain weight of coals upon which a *royalty* is paid: e. g. 10 tons at 1s. per ton.

WHEEL BRAE (S.). A flat or landing on the top of a *jig.*

WHEEL-HOUSE (B.). A shed for protecting the *horse-gin* on the surface.

WHIM. A *winding* (1) *drum,* &c., worked by a horse.

WHIMSEY. An old word for the hoisting apparatus at a mine, now known as the *winding engine*, which see.

WHIN. 1. A very hard, compact, dark-coloured, intrusive, igneous rock, composed of about 50 per cent. of silica, and having a sp. gr. of about 3, with a dull conchoidal fracture.

2. (S. N.) Any very hard resisting rock coming in the way of miners.

WHIN DYKE. A *fault* or fissure filled with *whin* and the débris of other rocks, sometimes accompanied by a dislocation of the strata. The Cockfield Fell Whin Dyke is probably the largest in Great Britain. It runs in almost a straight line, from near Carlisle on the west, to the east coast a few miles south of Whitby in Yorkshire. Whin dykes attain a thickness of as much as 200 feet in some places. See also *Dyke, Trap*.

WHIN-FLOAT (S.). A kind of greenstone, basalt, or trap, occurring in coal measures.

WHIN GAW (S.). Synonymous with *Whin Dyke*.

WHINSTONE (N.). See *Whin* (1 and 2).

WHIPSY-DERRY. See *Derrick*.

WHITE-DAMP. Carbonic oxide (O. 57 C. 43). A gas occasionally met with in coal-mines, which, although it will support combustion and is inflammable, quickly destroys life.

WHITE ROCK (S. S.). Intrusive *dykes* of Doleritic rocks in the coal measures: in external appearance it closely resembles sandstone.

WHOLE or WHOLE MINE (N.). That portion of a coal seam being worked by driving *headings* into it only, or the state of the mine before *bringing back* the *pillars*, or what is called *working the broken*, commences. See *Barrier System* (Fig. 10); also see *First Working*.

WHOLE CRADLE (N.). A platform or scaffold of nearly the same diameter as the *pit-shaft*, and hung upon chains attached to a crab-rope from the surface.

WHOLE FLAT (N.). A *panel* or *district* of *whole*.

WHOLE STALLS (S. W.). Two or more *stalls* having their *faces* in line or on a *thread* with one another.

WHURR. The buzzing noise made by the vanes of a *fan*.

WICHET (N. W.). A *working place* in the shape of a wide *heading* or *board* (1), sometimes 60 or 70 feet in width.

WICKET (N. W.). See *Wichet*.

WICKET WORK (N. W.). A kind of *pillar and stall* system of working a seam of coal, with *pillars* up to 15 yards and *stalls* up to 24 yards wide. A *plan* (2) of this description of workings would much resemble Fig. 113 (see *Single Road Stall*), the chief difference being that two roadways are generally carried up each *wicket*.

WIDE WORK (Y.). A South Yorkshire system (now nearly obsolete) of working coal. Sets of short *stalls* or *banks* (4), 7 or 8 yards in width, forming a line of *faces* about 60 yards, were carried to the *rise*, about 3 or 4 feet of coal being left between each *bank*, the main road pillars being subsequently extracted. See Plan, Fig. 137.

Fig. 137.

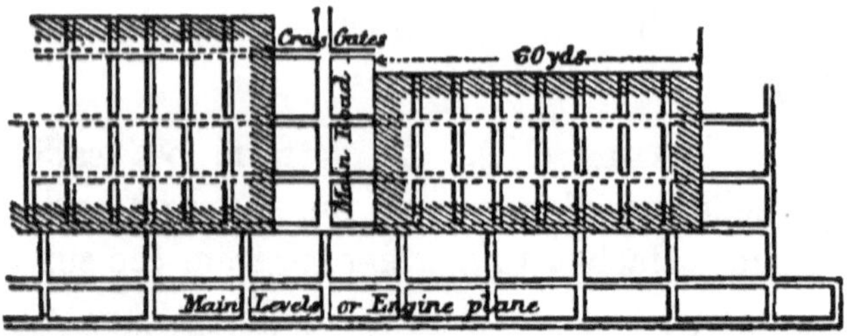

WILD-FIRE. An old term used by colliers for *fire-damp*.

WILD GROUND, WILD MEASURES, WILD STUFF (S.S., Sh.).

WIMBLE (N.). A kind of auger and scoop combined, for extracting the débris from *bore-holes* (1).

WIN. 1. To *sink* a *shaft* or *drive* a *drift* to a *workable seam* of coal, ironstone, &c., in such a manner as to enable you to effectually prosecute the working of it; or for the purpose of opening out a *district* in a mine, which, previously to *winning* the mineral, was cut off by a *fault* or by some other barrier.

2. (S.) *Won, found, proved*, (1) *tapped*, (2) *sunk to*, &c.

WINCH. A kind of windlass or crab for coiling ropes upon.

WIND. 1. A hand-windlass or *jack-roll*.
2. The atmospheric air circulating in a mine.
3. To raise coals, &c., by means of a *winding-engine*.

4. A steam-engine used purposely for lowering and raising men in an *engine pit* or pumping-shaft.

5. A single journey of a *cage* from top to bottom of a shaft, or *vice versa*.

WINDBORE. See *Sliding Windbore*, but made without the inner telescopic arrangement.

WIND-GAUGE. An *anemometer* for testing the velocity of the *wind* (2) in mines.

WINDING. 1. The operation of raising by means of a steam-engine, with ropes and *cages*, the produce of the mine.

2. (M.) Any underground road used expressly for ventilating purposes.

WINDING ENGINE. The apparatus fixed within a few yards of the mouth of a *shaft* for raising the minerals from the bottom, or from various levels, to pit top. It usually takes the form of a steam-engine, which first came into use for this purpose about the year 1763 at Hartley Colliery.

The modern *winding engine* consists of a pair of steam cylinders of equal diameter and stroke, placed either vertically or horizontal, the connecting rods being coupled direct through cranks at right angles to the main shaft, upon which the *drum* (1) is constructed, and which also carries the brake rim.

The following table gives the principal dimensions, particulars of work performed by, and other statistics in connection with a few of the most powerful winding appliances in the world:—

# A GLOSSARY OF TERMS

## TABLE OF SOME OF THE CHIEF PARTICULARS OF SEVERAL OF THE MOST POWERFUL WINDING ENGINES IN USE.

| No. | Name and Situation of Colliery, &c. | Type of Engine. | Cylinders. Dia. | Cylinders. Stroke. | Drum. Type. | Drum. Diameter. | Ropes. Size. | Ropes. Weight. | Load and Cage. Weight. | Time occupied in Drawing and Changing. | Steam in Boilers. | Depth or Lift. |
|---|---|---|---|---|---|---|---|---|---|---|---|---|
|  |  |  | ins. | feet. |  | feet. | ins. | cwt. | tons. cwt. | seconds. | lbs. | yards. |
| 1* | Moss Pits, near Wigan | Horizontal | 40 | 7 | Cylindrical | 25· | 4¾ c. | 50 | 5  0 | 65 + 45 | 40 ? | 711 |
| 2† | Harris' DeepNav., S. Wales | Vertical, inverted | 54 | 7 | Conical .. | 18 to 32 | 5¼ c. | 100 | 6  5 | 90 | 60 | 700 |
| 3‡ | Lumpsey Iron Mines, Cleveland | Horizontal | 42 | 6 | Conical .. | 17–21 | 5 c. | 20 | 7  12 | 30 + 25 | .. | 195 |
| 4§ | Ashton Moss, Manchester | Horizontal | 36 | 5 | Flat rope | 13–19 | 4¼ × 1 1/16 | 137 | 6  3 | 95 + 30 | 60 | 935 |
| 5‖ | Silksworth, Co. Durham .. | Horizontal | 48 | 6 | Conical .. | 15–28 | 5½ c. | 73 | 9  6 | 50 + 25 | 45 | 600 |
| 6¶ | Monkwearmouth, Co. Durham .. | Vertical, condensing | 68 | 7 | Flat rope | 22–25 | 5¼ × ⅞ | 95 | 7  8 | 124 | 25 ? | 580 |
| 7** | Sacre Madame, Belgium | Horizontal | 41⅜ | 5·25 | Flat rope | 5–10·4 | Taper | 125 | 4  4 | 65 + ? | .. | 765 |
| 8†† | Merthyr Vale, S. Wales | Horizontal | 32 | 4 | Conical .. | 6·5–24·5 | 4⅜ | 55 | 6  0 | 40 + 20 | 40 | 430 |

\* Stated to be 1200 horse-power.
† Load and cage to be increased to 10 tons, 10 cwt.
‡ To raise 1500 tons of stone in eight hours.
§ To be fitted with condensing apparatus.
‖ Fitted with cut-off gear.
¶ One cylinder. Balance chains used.
\*\* Brake wheel 16 feet 5 inches diameter.
†† These were originally marine engines.

USED IN COAL MINING, ETC. 291

Most large winding engines are fitted with steam brakes, some also with steam or hydraulic reversing gear, and with automatic cut-off or steam regulating gear. See *Water Balance, Köepe System, Drum* (1), *Conical Drum.*

WINDING ROPES. The ropes by which a *cage, chair, bowk, kibble, trunk* (3), &c., are raised and lowered in a *pit-shaft.* They are constructed of three different materials, viz. steel, iron, and hemp or manilla, and in two forms—round and flat. The former are sometimes made taper when of great length, the thicker end being of course that nearest or fastened to the *drum* (1).

The best quality of steel-wire rope, known as *plough* quality, costs about 5*l*. per cwt. Referring to the table of *winding engines,* above, it will be seen that in Nos. 4 and 7 instances the weight of the *winding rope* is in excess of the load (cage, tubs, and mineral) raised.

WINDING SHAFT OR PIT. The *pit-shaft* used chiefly for *winding* (1) purposes.

WIND METHOD. That system of separating coal into various sizes, and extracting the *dirt* (1) from it, which in principle depends upon the specific gravity or size of the coal, &c., and the strength of the current of air directed upon it, which is employed to effect such separation.

WIND ROAD. See *Winding* (2).

WIND WAY. See *Winding* (2).

WING-BORE (S.). A side or flank *bore-hole* (3).

WINNING. A *sinking pit,* a new coal, ironstone, clay, shale, or other mine of stratified minerals.

WINNING HEADWAYS (N.). *Heads* (1) driven in the *coal seam* at right angles to *drifts* (4).

WIRE (W.). A hauling rope.

WISKET (L.). A light basket, weighing about 25 lbs., used for carrying coals, &c., up a *shaft*.

WITCHET (N. W.). See *Wichet*.

WON. In mining language means proved, sunk to, and tested. Coal is *won* when it is proved and a position attained so that it can be worked and conveyed to *bank* (1). Coal may be *won* either by *levels*, by *drifts*, by *headings* to the *rise*, or by *headings* to the *deep*.

WOOD. Signifies *pit-props, bars, sprags, chocks, lagging*, &c., which are all used in various ways for supporting the *roof* and sides of underground *workings* and *ways*. The cost of *wooding* or *timbering* in a colliery ranges from say 2*d.* to 10*d.* or 1*s.* per ton, according as the *roof* is a good or a bad one.

The most suitable kinds of wood for mining purposes are :—

Fig. 138.

    For *props*, yellow or Norway pine.
    „ *bars*, larch, ash, elm, and fir.
    „ *sprags*, ash and fir.
    „ *chocks*, any hard and tough wood.
    „ *lagging*, any tough and durable odds and ends.

WOOD CHAIN (S. S.). A chain used for raising the minerals up the *pit-shaft*, composed of five links of iron in width, with small blocks of wood filling up the spaces in the links. See sketch, Fig. 138.

WOOD COAL. See *Board Coal*.

USED IN COAL MINING, ETC. 293

WOODERS (Y.). See *Timberers*.

WOOD RINGER. See *Ringer* and *Dog and Chain*.

WORK (M.). 1. A *stall* or *working place*.

2. Meaning *get* (2), in the sense of whether a coal gets or *works* easily or with difficulty.

3. When during the operation of *holing* or cutting coal a crackling or bursting sound is caused, the coal is said to *work*. Also when the *roof* shows signs of giving way, and cracks with a noise, it is said to *work*.

4. To carry on the various operations connected with the mining of coal, &c.

5. To get, cut away, or excavate and remove any bed or seam, or part thereof, of coal, ironstone, or other mine, whether underground or in *open work*.

6. (S. S.) A *side of work*.

WORKABLE. 1. A seam of coal is generally called a *workable* coal when (if of good quality) its thickness exceeds 18 or 20 inches. It may perhaps also be said that all mines of coal, &c., to a depth of 4000 feet, are workable.

2. Any seam or *rake* of ironstone that can be profitably mined.

WORK BOX (Lei.). See *Box*.

WORKED OUT. A bed of coal, &c., a pit, or a *lift* (10), is called *worked out* when all the available mineral has been extracted.

WORKING BARREL. The *pump tree* or cylinder in which the *bucket* moves up and down. It is usual to make it a little less in diameter than the ordinary pipes or *trees* (1). It is bored out in a lathe, and if the water to be pumped is very corrosive or ochrey, is lined with brass.

WORKING BEAM. See *Brake-staff*.

WORKING COST. The cost per ton of producing coal, &c., and loading it into wagons, boats, &c. It includes all expenses in *getting, haulage, banking, surface labour, management, sales, timber, stores, royalties, way leaves, rates and taxes, insurance, colliery consumption, bad debts, loss in wagons and stocks, repairs, &c., interest on capital, replacement of machinery*, &c.

WORKING FACE. See *Face* (1).

WORKING FURNACE. A *furnace* supplied with fresh air from the *downcast* pit.

WORKING HOME. Getting or working out a seam of coal, &c., from the boundary or far end of the *pit* (2) towards the *pit bottom*, thus leaving behind all *goaves, fire-stinks*, &c.

WORKING ON AIR. When the holes in a *snore-piece* are not completely covered with water, and air is sucked up with the water, the pumps are said to be *working on air*.

WORKING PLACE. The actual place in a mine at which the working of the coal, &c. [either by driving headings or by *stall work*], is going on : viz. a *head end* or at a *working face*.

WORKING OUT. Getting coal, &c., from the *shafts* outwards, or in the direction of the boundary of the colliery. The opposite to *working home*.

WORKINGS. 1. The portions of a seam of coal, &c. worked away, which, of course, includes all *roads, ways*, *levels, dips, airways,* &c., whether in use or not, together with the *stalls, headings, goaves, staples,* &c. The deepest coal workings in existence are said to be 3511 feet—at Gilly Colliery, in Belgium.

2. The quantity, tonnage, or output of minerals during a certain period from a certain lease, or a *district* in a pit. See *Get* (2).

WORM or WORM COIL. A tool, something similar to a *wad hook*, used for loosening tough clays at the bottom of *bore-holes* (2). See *Wad Coil*.

WREATHS (Lei). Four short pieces of hemp rope placed round the legs of a horse or pony and fastened together above its back, by which it was formerly lowered into or brought up out of a *pit-shaft*.

WRECK. See *Bore-meal*.

WRENCH. See *Key*.

WROUGHT. Coal, &c., worked or *gotten*.

WYE (C.). The beam-end connection above the pump-rods of a *winding* and *pumping engine*.

# Y.

YARDAGE. Cutting coal, &c., by the yard or fathom. In many districts a price per ton on the coals is paid, in addition to so much per yard.

YARD-STICK. An ash walking-stick, 3 feet in length (having a notch or other mark put upon it at every foot), which a *manager* or *underviewer* carries with him in the pit, with which he roughly measures any lengths of work done and other distances whenever occasion arises, and with which he chastises unruly lads.

YARD WORK (F. D.). Synonymous with *yardage*.

YARK (D.). To jerk a rope or other appliance used for lifting or drawing.

YED (Lei.). See *Head* (1).

YIELD. 1. *Pillars* of coal are said to *yield* when they commence to give way or crush.

2. The proportion of a coal seam, &c., actually sent to *bank* (1).

YOKES. Short sawn timbers placed across *biats* for steadying *pump trees*. See *Chogs*, Fig. 40.

# Z.

ZONE. In coal-mining phraseology, this word signifies a certain series of coal *seams*, with their accompanying shales, &c., which contain, for example, much *fire-damp*, called a *fiery zone*, or, if much *water*, a *watery zone*. As a rule, the *fiery zone* begins immediately below the upper or *water-zone*, which does not usually descend below (say) 600 feet.

www.ingramcontent.com/pod-product-compliance
Lightning Source LLC
Chambersburg PA
CBHW022026240426
43667CB00042B/1199